PHYSICAL GEOGRAPHY OF WESTERN TIBET

PHYSICAL GEOGRAPHY
OF
WESTERN TIBET

H. STRACHEY

GYAN

Published by

GYAN

Gyan Publishing House
5, Ansari Road
Daryaganj, New Delhi-110002
Phone: 011-47034999, 9811692060
E-mail: books@gyanbooks.com

Distribution Network
gyanbooks.com
India, USA, Canada, UK, Australia, France

ISBN : 978-81-212-3889-2 (PB)
First Published, London, 1854

2nd Impression 2020

Printed at: Gyan Press, Delhi.

Physical Geography of Western Tibet
Author: H. Strachey

PHYSICAL GEOGRAPHY

OF

WESTERN TIBET.

BY

CAPT. H. STRACHEY, F.R.G.S.

LONDON:
PRINTED BY WILLIAM CLOWES AND SONS, STAMFORD STREET, AND CHARING CROSS.
1854.

PREFACE.

The following Memoir contains the purely geographical part of a report submitted to the Indian Government after my return from the Tibetan Boundary Commission, which was deputed to Ladak by the Governor-General (Lord Hardinge) in 1847. The diplomatic object of this commission was to define officially the territorial boundary between the trans-Himalayan possessions of our new ally and dependent Maharaja Gulab Singh, and the Tibetan provinces subject to China; but the Chinese-Tibetan authorities declining, as usual, to hold any intercourse with the British, or to admit us within their territory, our commission, consisting of Major A. Cunningham, Dr. T. Thomson, and myself, was admitted only through the British Hill Provinces into Ladak, and left to employ itself during the two years between the summers of 1847 and 1849 in such general geographical explorations as were still open to us. The barbarous anarchy of the mountain tribes on the N.W. frontier beyond the Indus offering as great an obstacle to travelling on that side as the jealousy of the Chinese Government on the E., we found ourselves confined to Ladak and Balti, and the neighbouring Himalayan provinces in the hands of Gulab Singh, regions already visited and described by Moorcroft and Trebeck, and Mr. Vigne, so that we could only collect further information in the same field, without effecting much new exploration.

My own attention was chiefly directed to the improvement of the Map of West Tibet, of which an excellent outline had been already constructed by Mr. Arrowsmith from the field books of Trebeck; the political boundary which the Chinese Government refused to adjudicate politically was also ascertained and laid down from my own local examination; and in the course of this work I collected what subsidiary information I could upon the geography, statistics, and political condition of the country. My personal knowledge of the country inside the Chinese border was obtained by some small expeditions made "upon my own responsibility," without the cognizance of either government, once in the autumn of 1846, from British Kumaon into the S.E. corner of Nari-Khorsum (a separate account of which was published in the Journal of the Asiatic Society at Calcutta), and again in the summer of 1849, on my return from Ladak into British Garhwal, through the S.W. end of the same province, and one or two others of less note.

The Memoir now published is confined to the purely physical geography of West Tibet. To be complete as such it should have included a notice of the scientific geology, botany, and zoology of the country; but I have not sufficient knowledge of these subjects to discuss them properly, and the botany has been fully treated by my colleague Dr. Thomson, to whom that department was specially assigned; Major Cunningham has also written a general account of Ladak, which is about to be published by the Indian Government. A quantity of statistical and political information, collected by myself, which was included in my official report, but is less suited to the Geographical Journal, will perhaps see the light otherwise at a future opportunity.

The Map accompanying this Memoir has been constructed by Mr. Arrowsmith from the materials furnished by myself. Another, on a much larger scale, to show all the details of my own explorations, and include everything else that I have been able to collect of the topography of these countries, is in course of preparation by myself, and will be published, if the means be forthcoming, some time this year.

The heights above the sea level given in the Map and Memoir are from a large collection (many hundred) of observations by myself and other travellers (particularly Dr. Thomson, Mr. J. E. Winterbottom, and my brother Richard), many of those on the

S. side of the mountains being trigonometric and barometric determinations, but on the Tibetan table-land mostly by boiling thermometer. The majority of these last were made by myself with considerable care, and all have been computed by myself from the original data; the measurements being checked by comparisons with heights geometrically or barometrically determined, approximate probably within one or two hundred feet of the truth.

My statements of the lengths of river courses and the like are derived from opisometer measurements on my own maps, adding a seventh for the probable excess of minor sinuosities, which is so moderate an allowance, I believe, as to preclude all exaggeration.

The orthography of proper names follows the system of Sir William Jones, as adopted by the Asiatic Society of Bengal, which may be succinctly explained to consist of English consonants with Italian vowels. I have simplified many of the Tibetan names, however, to render them intelligible to an English eye, but if read by the above rule they will approach nearly to their proper pronunciation.

HENRY STRACHEY.

London, January, 1854.

CONTENTS.

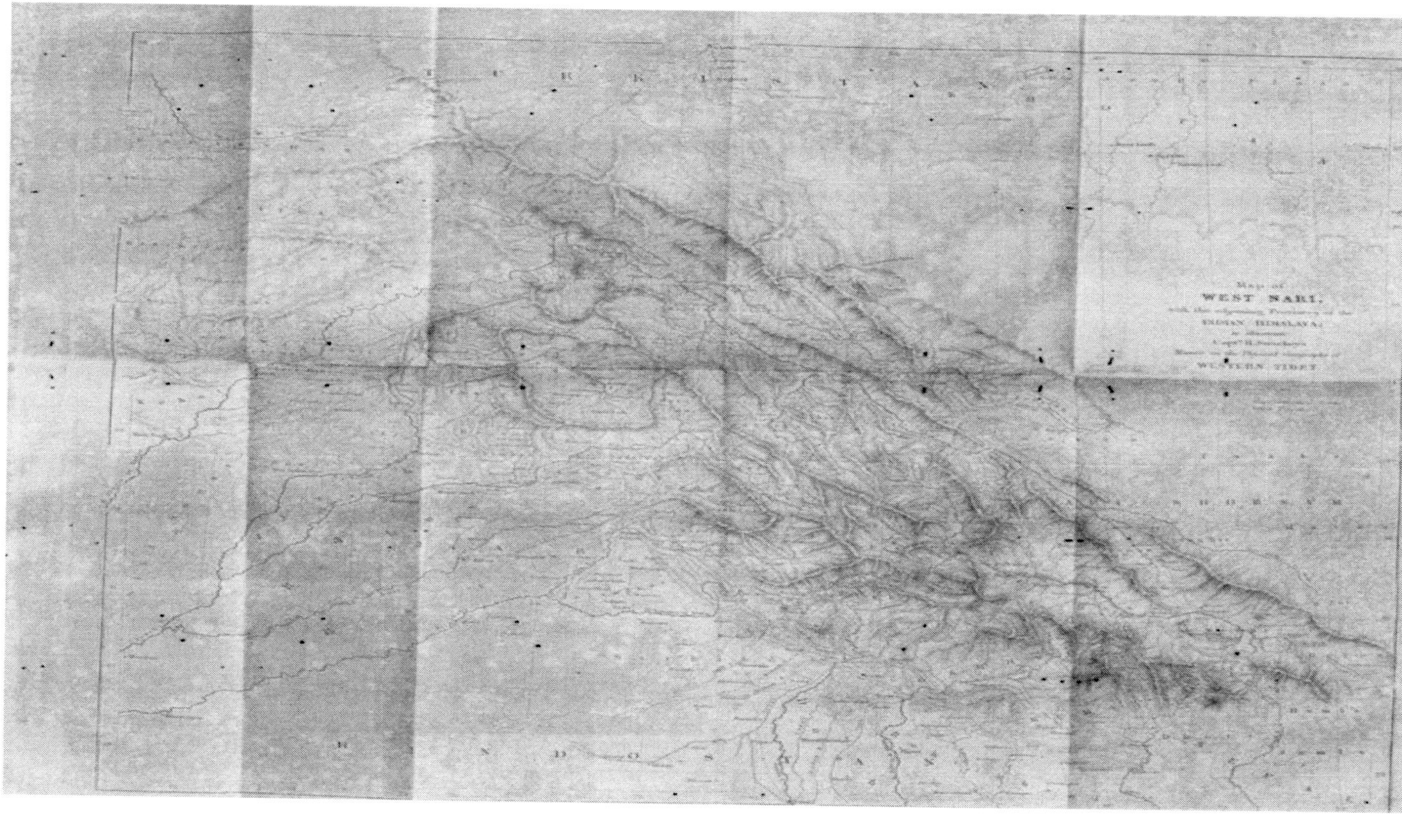
Map of
WEST NARI,
INDIAN HIMALAYA,
WESTERN TIBET

PHYSICAL GEOGRAPHY

OF

WESTERN TIBET

Tibet.—As my purpose in the following pages is to confine myself generally, if not rigorously, to the result of my own observations and inquiries, I shall say nothing more of Tibet in general than may serve as a starting point to these, and explain the names and allusions which must often occur in them.

Tibet is an extraneous appellation unknown to the inhabitants of the country we so designate, and apparently adopted by Europeans through the medium of the Asiatic Turks and Persians, whose *Tbt* some English writers have improved into *Thibet*, and French and Germans into *Tubet.* I cannot set aside a word which long prescription has so completely naturalized in our language, but in using it I shall retain the simplest form and the nearest to the Turkish original. The true autochthonous name of our Tibet is *Bodyul,* which would be properly rendered into English *Bodeland:* it is also called simply *Bod.* The *Bhot* (and *Bhotiya*) of the Indians is no doubt derived from this word, although the Tibetans have stories otherwise accounting for the Indian name. The *Hun* and *Hundes* of the Hill Indians are about equivalent to our *Tartars* and *Tartary;* and where these are used, the more correct terms are improperly restricted to those Tibetan stragglers who have crossed the Indian passes, and applied to the Himalayan regions in which they have taken up their abode.

The Tibetans, no less than other nations, greatly corrupt the indigenous names, or invent spurious names of their own, for the foreign countries nearest or best known to them; and the geographer should understand these. *Gyanak, i. e.* the *Great Black,* is *China,* so called from the predominance of black (or dark blue) in the dress of the Chinese people: but a Chinaman is called simply *Gyami, i. e. Great Man,* and in Tibetan estimation he is pre-eminently such. *Gyagar, i. e.* the *Great White,* is *India,* so called from the general dress of the people. *Gyaser, i. e.* the

Great Yellow, is *Russia*; whether so called from the supposed prevalence of auburn hair or yellow clothing among the Russian people, I know not. *Filing* is a corruption of the Persian and Turkish *Farang*, denoting *Europe*: *Gyafiling*, *i. e. Great Frankland*, though properly signifying *Europe*, is commonly applied to *British India*. *Horyul* is *Turkland*—in the language of its own inhabitants *Turkistan*; and particularly the South-eastern region next to Tibet, or the Kingdom of Kashghar, which some European geographers have called—most improperly—*Little Bucharia*. *Sok-yul* is the land of the *Sokpo*, whom Europeans call *Mongol*, and the Turks *Kilmak* (both no doubt incorrect substitutions of a part for the whole). Such terms as *Khamsok* and *Horsok* are sometimes applied to the countries between East Tibet, S.E. Turkistan, and Mongolia; denoting a mixture, whether real or imaginary, of Tibetans and Mongols, and of Turks and Mongols, where those countries border upon each other.

The Indian Provinces next adjoining to Tibet are *Ashong*, *i. e. Acham*, or *Assam*. *Monyul*, the whole of the *Indian Himalaya*, especially the *Sub-Himalaya* inhabited by *Mon*, *i. e. Hill Indians*. *Lhopato*, *Lho-duk*, or *Lho-mon*, the *Bhotant* of Bengal, or *Bootan* of the English. Its capital *bKrashismchhosdzong*, or pronounceably *Tashichuzong*, *i. e.* the *August City of Religion* (improved by the English to *Tassisudon* and the like). *Demojongs*, *i. e.* the *Goodly Region*, the *Shikim*, *Shikimpati*, or *Sikkim*, of Indians and English: the British cantonment in its lower part is *rDorjegling* or *Dorjeling* (corrupted to *Darjeeling* and the like), a name equivalent to *Vajrapura*, or in plain Saxon *Thunderbolton* (otherwise the *Place of the Heavenly Sceptre*, or of the *Precious Stone* or *Diamond*): *Kangschan hJing*, *i. e.* the Icy Mass, is the great snowy mountain on the N. border of this province, famous as the highest measured peak on the globe. *Palbo* is *Naipal*; *Kyunam*, *Kumaon*; *Galdiya*, *Garhwal*; *Chongsa*, *Himalayan Garhwal*; *Kunu* is the District of *Knor*, *Kanor*, *Kanoring*, *Kanaur*, *Kunawar*, &c.; *Nyungti*, *Kullu*; *Garzha*, *Lahaul*. *Panga* of *Himalayan Chamba*, *Paldar* and *Maru-Wardwan* of Himalayan *Kishtwar*, are called by their native names. *Kacheyul* (vulgo *Kachul*) is *Kashmir*.

The only part of Tibet with which the English have any direct acquaintance is that which borders upon the north-eastern confines of British India. This region extending from the mountain gorge of the *Brahmaputra* to that of the Indus, is probably little more than the western half of all Bodeland; for the eastern region extends to the western frontier of China, and to a greater breadth than the western, and the Tibetans themselves designate the latter sometimes *Bodchan*, *i. e.* Great Tibet, but its proper name is *Kham* or *Khamyul*; and I have no further concern with it for the present. For the exploration of East Tibet we are indebted to the French missionaries Messrs. Huc and Gabet, who, in 1846-7, traversed the whole breadth of *Khamsok* on their way from North China to the capital of Central Tibet, and of *Khamyul* on their return to Central China ; and Mr. Huc's well known narrative of this most adventurous journey contains, I believe, the sole direct European information regarding those parts of Tibet.

The region known to us as lying behind the Indian *Himalaya* is a belt of high mountainous table land, narrow compared with its length, and (to the best of our imperfect knowledge) subsiding, on its N.E. border, into the plains and sandy deserts of *Turkistan* and *Khamsok*. This part of Tibet is called *Bod, i. e. Tibet proper*, or Central Tibet, at its E. end, and *Nari* at its N.W., the former division being the shorter of the two, but probably the broader, and certainly the more populous and civilized.

The line separating *Bod* from *Nari* may be taken approximately as a continuation of that dividing the E. end of *Naipal* from the British dependency, or annexure, of *Demojong*, which latter is also the demarcation of Indian races from those of the Tibetan family on the S. slope of the *Himalaya* (or more strictly, of the predominance of those races in the *Sub-Himalaya*). The division of *Nari* from *Bod* is partly natural, inasmuch as the elevation of the country becomes inimical to agriculture and population to the westward, and lower, warmer, more habitable, and more cultivable to the eastward. But there is no great geographical landmark that we know of, nor any ethnical or political separation of the inhabitants.

Bod proper is subdivided into several territories, provinces for extent, though mere districts for population. These are (from E. to W.) *Kongbo, Takpo, dvUs*, and *gTsang ;* and perhaps others. *Kongbo* being so warm as to admit the general cultivation of rice, and so moist as to be well wooded with natural forest, may be reckoned with probability to lie about the valley of the *Brahmaputra* immediately before its downward passage to the noxious climate and barbarous people that oppose the ascent of the English explorer from *Ashong*.

In *dvUs*, already elevated into a cold climate, lies the *Gyalsa* (*i. e.* Capital) of all Bodeland, the City of *Lhasa* (*i. e. God's-ground*), the *Rome* of Tibet; which has been rescued from an almost mythical obscurity by the enterprising French travellers above mentioned. Of *Tsang* and its monastic capital *Tashi Lhunpo* (*the August Lump*) the English reader may find a faint, dreamy sketch in the pleasing narrative but vague geography of Turner. These two provinces are often coupled together in Tibetan phraseology under the name of *Utsang* (merely an abbreviated compound of the two names), which may be taken to signify generally *Central Bod*, as distinguished from *Nari* and *Kham*.

Lhopato and *Demojong* being inhabited by Tibetans retaining their own language, manners, and governments (more or less modified by Indian influences), would be included by a Tibetan geographer in a complete account of his own country; but in physical geography they belong rather to India. Regarding the former of these provinces information may be sought in the pages of Turner and Pemberton; and the latter has recently been laid open to the arts and arms of *Dorjeling*.

Nari, or Western Tibet.—*Nari* is divided by the Tibetans into three great provinces, viz., *Mangyul*, *Khorsum*, and *Maryul*. The first of these is nearly conterminous with the present kingdom of *Naipal*, but not extending quite so far to the W. It is probably as broad as that country, or broader, with near an equal area, but not a hundredth part of its natural and political resources, being so elevated, cold, and dry as to support little beyond a scanty population of shepherds with a few small villages in the warmest parts. The chief districts of this province are (reckoning from E. to W.) *Tingri* (*Lawn Mountain*), *Shelkhar* (*Glass Castle*), *Nyanam*, *Khyirong* (*Dog Valley*), *Chamshen Tsuglakang* (containing a Tibetan *University !*), and *Kungtang*, on the southern border; and some of them forming the heads of valleys that enter *Naipal*, and containing agricultural villages: *Zangzang*, *Kyado*, *Semukul*, *Saka*, *Nyugu*, *Tradum* (a monastery), *Troshot*, and *Shamtsang* are in the central part, all high pastoral upland; and the *Dzong*, or seat of local government of the whole province, is at *Saka* (otherwise a mere shepherds' hamlet): *Beng Madma* in the N.W. quarter, also pastoral upland, though belonging geographically to *Mangyul*, is, and has been for many centuries, included in the government of *Nari-Khorsum*. I know little beyond the names and *relative* position of all these places; nor can we expect to learn much of *Nari-Mangyul* whilst *Naipal* itself is debarred to the English traveller.

Nari-Khorsum is that N.W. extremity of the Chinese Empire (but not the westernmost, for *Kashgar* extends to the meridian of *Peshawr*) with which the British have been in contact since they annexed the *Himalayan* provinces between the *Sutluj* and *Kali*; and *Nari-Maryul*, the N.W. extremity of all Tibet, comprising the modern provinces of *Ladah* and *Balti*, already well known by the descriptions of Moorcroft and Mr. Vigne. When I use the term *West Nari*, it may be understood to mean all *Nari-Khorsum* and *Maryul*, to the exclusion of *Mangyul*, and my own explorations are confined to this region.

I leave the Indian confines of Western Tibet to the Indian geographer; and most parts of them, W. of *Naipal*, have been explored and described in detail

With a small exception, we know very little regarding the

countries that bound it on the N.W. and N.E. In the former direction it seems pretty certain that the great mountainous mass of the Tibetan table-land and Indian *Himalaya* is continued much in the same style till it ends abruptly about the 40th parallel of N. latitude, in the mountains which the Turks call *Bulut Tag* (*i. e. Cloud Mountains*), E. of *Samarkand*, and S. of *Khokand*, in the very centre of Asiatic *Turkistan*. This name is almost as significant of great height as the *Himalaya* (*Abode of Snow*) of the Indians; and the blank in civilized habitation, the separation of empires, and the scanty commercial communication in that quarter, seem to point to the same conclusion; though no actual account of the region that I know of is to be found in any of the published Geographies. The former direction of the chain is perhaps continued beyond the *Bulut Tag* towards the *Aral* Sea, but if so, in mere hillocks compared with the mountains of *Pamir* and *Nari*. In elevation and mass, however, these mountains have a decided continuation, with a very acute change of direction, in the great chain which the Chinese geographers call *Thian-Shan* (*i. e. Celestial Mountains*) projected to the eastward, through the eastern extremity of *Turkistan*. The *Afghan* range, which continues the watershed of the Indian regions to the westward, can only be looked upon as a long spur or branch much inferior in elevation to the main trunk, the high masses of the *Hindu Kush* being as it were the point of articulation.

Mr. Wood's description of *Badakhshan* and *Pamir* presents a remarkable likeness to a province of the Indian *Himalaya* (such as *Kanor*), communicating by a valley gorge (as that of *Tsotso*) with a Tibetan upland (like *Rupshu*). On both the summits we have 15,000 feet lakes embedded in 19,000 feet mountains, with the same zoology of domestic yak and wild sheep; and the *Khirgiz* even is cousin-german to the *Changpa* of *Nari*.

The countries occupying the main mountain mass between *Pamir* and the N.W. extremity of *Maryul* (including *Kafiristan*, *Chitral*, *Yasen*, &c.) have been so completely barred against European research by the barbarism of their inhabitants, that their names even are uncertain, and we know scarce anything of their geography beyond the fact of their being highly mountainous; but the turbulent and predatory character of the people would prove them to be ethnically related to the Afghans rather than the Western Tibetans; and as they exhibit themselves in their national rôle of gang-robbers up to *Gilgit*, on the frontier of *Balti*, I shall consider Tibet to be terminated there by the southward turn of the Indus towards India, and by the lofty spurs of mountain which project to the northern point of the river bight from a Turkish watershed, dividing the north-western extremity of Balti from the non-Tibetan countries of *Nagar*, *Hunz*, *Kanjut*, and *Gilgit*.

E. from this point the northern confines of Tibet are better determined, as far E. as the meridian of 82° or 83°. Native information from the side of India and Tibet (the best of it collected by Moorcroft), coupled with the Chinese geography imparted to Europe by the savans of Paris and Berlin, and the very slender accounts of one or two antique European travellers, assure us that the mountainous table-land descends in that direction, by hilly slopes similar to the *Sub-Himalaya*, to a great plain like that of India, extending indefinitely to the eastward, and inhabited by true Asiatic Turks, of late years under Chinese government. This northern mountain slope is almost uninhabited, and in its upper part naturally very barren from elevation, cold, and excessive dryness of climate; but the plain below is well cultivated by a civilized population, excepting its S.E. quarter, where scarcity of water and abundance of sand convert it into an impracticable desert, apparently the S.W. extremity of the great *Shamo* or *Cobi* of the Chinese geographers.

The Chinese-Turkish provinces lying next to the mountain foot are *Kashgar*, at the N.W. end, in the corner between the *Bulut Tag* and the *Thian Shan; Yarkend*, below *Pamir Kanjut* &c., and *Balti;* and *Khotan*, under *Ladak* and *Nari-Khorsum.* The communications between *Balti* and *Yarkend* are now almost closed by political barbarisms, superadded to natural difficulties, and the only extant intercourse between the Turkish and Tibetan countries confined to one oblique line between *Ladak* and *Yarkend.* The direct passages from *Nari-Khorsum* to *Khotan* are totally unused, but I am well assured of their existence, and in the lower part of the mountain slope lies the fertile and populous district of *Serikia*, through which there was once, according to Moorcroft, a high road from *Gar* to *Khotan*."

From information of my own, as well as from Klaproth's map, I gather that the province of *Khotan* lies further S. than *Yarkend*, and consequently that it can admit of no great expansion of the Tibetan table-land up to the meridian of 82° or 83°; but E. of that my information of the northerly confines of Tibet becomes as vague as our European versions of the Chinese maps; yet it seems certain that the *civilized* regions of *Nari-Mangyul* and *U-Tsang* are still confined to a narrow belt. The inhabitants and travellers of this zone complain of the incursions of a lawless tribe who occupy their northern borders, and are said to call themselves *Rundur*, though by the civilized Tibetans commonly termed *Kyampo*, i. e. *Nomads*, which is vulgarly corrupted into *Khampa*, and so improperly confounds these demi-savages with the peaceful and civilized people of Eastern Tibet. The *Rundur* are said to be essentially Tibetan in their physique and language, but barbarous in dialect and manners, almost independent of the Chinese government in their own haunts, and visiting the civilized regions on their S. as much for plunder as for trade. The last stragglers of them come as far W. as *Gar* and *Ruduk* of *Nari-Khorsum*, but they chiefly infest the roads between *Nari* and *UTsang*, avoiding the government parties, but robbing private persons, sometimes killing, and even, it is said, *eating* them; the best of them, however, occasionally doing a little honest trade in shepherds' products. They dwell in tents only, roving much from place to place, and live exclusively on the flesh and milk either of their domestic cattle—horses, yak, and goats and sheep—or of the Tibetan antelope (*Tsos*), which they catch in snares or pitfalls (fire-arms being scarce among them); and corn or any other vegetable food is hardly known to them—they are, in short, complete *Tartars*. The country of the *Rundur* cannot be over accessible to the civilized Tibetans themselves, and they certainly know very little about it, although the Chinese geographers have pretty well sprinkled it with lakes, rivers, mountains, and names. For my own part I exhaust my knowledge of its natural geography by saying that it is elevated and mountainous, and apparently the northern part of the Tibetan table-land, bounded on its N. by the S.W. end of the Great Desert, which itself may be very elevated here, as the Russians have found some parts of it at the N.W. end. This desert is well known to extend as far as *Khotan*, and completely hems in that province on the E., leaving its only external communications towards the plain of *Yarkend* or the mountains of *Nari*.

General River System.—*Nari-Mangyul* is separated from *Nari-Khorsum* by a natural landmark, viz. a transverse mountain ridge running from the N. face of the Indian watershed, more or less to the N., across the breadth of the central upland, and itself constituting a great watershed that divides all *Nari* and *UTsang* into two main basins of drainage. The major axes of both these lie parallel to the longer direction of the table-land, till they attain the further extremities of *Bod* and *Nari* respectively, where they become deeply sunk, and turn rather abruptly through the *Himalaya* to enter the plains of India.

The river that carries the drainage of *Nari-Mangyul* and *UTsang* to the south-eastward is called by the Tibetans the *rTachok Tsangspo*, i. e. *Horse River*. The best of my Ladak informants could not assure me positively of its course below *Lhasa*, but assented fully to its identification with the main trunk of the *Brahmaputra* river, as asserted (and all but established) by the geographers of Bengal. The river which drains the greater part of *West Nari* to the north-westward, called by the Tibetans *Senge-Tsangspo*, i. e. *Lion River*, is now well established (by the explorations of Mr. Vigne, and subsequent surveys of Lieut. R. Young) as the chief source of the *Indus*—a fact which English geographers have had to rediscover for themselves within the last half century, though I find it distinctly stated in a book (and that not professing geography) written by a Tibetan monk 250 years ago.

But the *Indus* is not the only river of *West Nari*; all the S. part of *Nari-Khorsum*, and a small eastern corner of *Maryul*, are drained by the heads of the Indian *Satadru*, v. *Sutluj*, the chief of which, called by the Tibetans *Langchen Tsangspo*, i. e. *Elephant River*, originates in the southern part of the transverse watershed that divides *Nari-Khorsum* from *Mangyul* (the heads of the *Indus* occupying its northern part), and breaks through the *Indian Himalaya* in Upper *Kanor*.

If we conceive the whole breadth of Asia towards its medial meridian to be divided into three great zones, approximately parallel to the equator, of which the northernmost or Siberian drains into the Polar Seas, the central or Turkish into inland seas or sandy deserts, and the southern or Indian into the Indian Océan—the western half of Tibet will occupy the very northernmost border of the last region, forming in this respect an easterly continuation of *Afghanistan*, and the great S. Asiatic watershed will lie along the northernmost confines of the Tibetan table-land, dividing Tibetan from Turkish waters, and altogether behind the *Indian Himalaya*. The western prolongation of this watershed has been repeatedly crossed by English travellers in the hills of *Afghanistan*, but in all Tibet only just touched at a single point of *Maryul*; Dr. Thompson having in the summer of 1848 crested the *Karakorum* Pass, which divides a northern affluent of the Indus from a stream pretty well known to join the river of *Yarkend*, and Mr. Vigne in *Khapalu* of *Balti*, and myself in *Nubra* of *Ladak*, reached terminal glaciers, supposed to be upon the S. slope of the same watershed; but with these exceptions the Tibetan line is unexplored, and known only by meagre native reports.

Persons not familiar with mountain geography are apt to consider a main watershed as identical with a high mountain range, and the idea is of course founded on natural facts; but in detail the exceptions almost exceed the rule, and the chief watersheds will often be found to follow the lowest of the ridges—mere valleys even—and the channels of drainage to cross the highest; nor will this be thought strange if we regard the depth and number of the fissures that intersect these mountains, often directly transverse to the main lines of elevation, and the sufficiency of the slightest slopes to establish a flow of water no less positive than the steepest.

The line that I have termed the Indian watershed is not the crest of the great *Himalaya* as seen by the Indian observer, but a succession of valley heads much depressed, and penetrating that mass to such a depth (in *Kumaon* and *Garhwal*, according to my brother, 35 miles) that the passes from India to Tibet are never visible from any station fairly S. of the perpetual snow. These recesses are of various depths; the smaller and more numerous discharge the shorter sources of the Indian rivers, such as the *Vet* of *Kashmir*, the *Chandra-Bhaga* of the *Panjab Himalaya*, the *Jamna* and *Ganges* of *Hindostan*, and no doubt many tributaries of the latter flowing from *Naipal* to *Bengal;* but a few of them are so great as to form an absolute break through the outermost ranges of the snowy mountains, and make the sources of some of the Indian rivers absolutely Tibetan or *Trans-Himalayan*, such as the heads of the *Tsotso Sutluj*, the *Karnali Gogra Purang*, or *Mapcha* river (in S.E. *Nari-Khorsum*), one branch of the *Trisul-Gandaki* (the Tibetan *Khyirong*) and the *Aran-Kosi* (?) of *Naipal.* The passage of the *Langchen Sutluj* is intermediate, as it were, between such indentations and the two great penetrating fissures of the *Indus* and *Brahmaputra*. The northern or Turkish watershed of W. Tibet may possibly be indented and depressed in the same style; but we have no data for safe speculation on the subject.

The Tibetan table-land lying between the Indian and Turkish watersheds may be regarded as the flat top of a great embankment, exhibited in all its thickness in the scarp of the *Indian Himalaya;* the summit, though deeply corrugated with valleys and mountains in detail, being in its general relief laid out horizontally at a height little inferior to that of its southern scarp. The highest summits as yet known and measured certainly lie upon the Indian watershed (or even projected beyond it to the southward); and the abundance of large glaciers in the Turkish watershed of *Maryul*, in spite of the very dry climate of that region, with their general absence in the central parts of the table-land, is indicative of greater height in the former; yet very lofty summits are to be met with in all parts of the interior, some of which when measured may perhaps prove equal to the highest peaks of the *Indian Himalaya;* and the passes which must be crossed to get from one Tibetan valley to another, even in the very central axis of drainage, generally equal those by which Tibet is reached from India; so that on the whole, I think the medial depression is but faintly marked in the beds of the main rivers, without much affecting the mean elevation of the mass.

As none of our modern travellers have visited any part of the great transverse watershed that divides the heads of the *rTachok* from those of the *Langchen* and *Senge* rivers, or *Nari-Mangyul* from *Nari-Khorsum*, for any vague idea of its character we are left to our own conjectures and native report. In the former we must not lose sight of the general habitudes of the mountain watersheds above adverted to; but the latter leads us to suppose that the separation of the drainage (in its southern part, at least) is marked by a well-defined mountain ridge articulated to some great *Himalayan* mass, and running some way northward across the table-land. The high road from *West-Nari* to *U-Tsang* crosses such a ridge by several passes, of which the most frequented is called *Maryum La*, and this in Tibetan statements is commonly reckoned the boundary both natural and political, though essentially but one point in it, as the term *La* means an individual pass, and not a whole mountain range.

From this, north-westward, my geography becomes more definite, and embraces the results of my own personal explorations.

West-Nari in detail.—The general direction of *West-Nari* is S.E. and N.W. in its longest dimension—that is, *Maryum La* lies approximately in E. long. 82½° (or a degree E. from *Manasarowar*), and the southernmost point of *Nari-Khorsum*, on the Indian Himalaya of N.W. *Naipal*, reaches probably to N. lat. 30°. The southward bend of the *Indus*, where it leaves *Balti*, lies near the meridian of 75° E. long., and no part of Tibet seems to extend further W. Its northernmost extension at this end is probably about 36½° N. lat. The direct length of the major axis of this tract is near 600 miles in a straight line drawn from S.E. to N.W.: the undulated line of the Indian watershed, with all its sinuosities, perhaps exceeds 700.

The furthest personal explorations of English travellers at the S.E. end have extended only to the W. shore of *Manasarowar*, about long. 81° 20′ (Moorcroft and Hearsay, 1812; H. Strachey, 1846; J. E. Winterbottom and R. Strachey, 1848), and to the middle of *Purang*, in lat. 30° 15′ (H. Strachey, 1846); but we have obtained a distinct though distant view of the mountains in this quarter as far as E. long. 82°; and native information regarding the remaining corner to the S.E. has been tolerably precise. To the N.W., Mr. Vigne, in 1832, was the first to reach the furthest westing of the Tibetan Indus and of all Tibet, near *Acho* of Lower *Balti*, about E. long. 74° 50′; his furthest northing, in *Shigar* of the same province, was about N. lat. 35° 40′, still something short of the Turkish watershed. The subsequent explorations of Lieut. R. Young, with Messrs. J. E. Winterbottom and the late P. Vans Agnew, in 1847, extended a little further N., viz. to 35° 50′, in the small valley of *Haramosh*, at the north-western extremity of *Balti*. In the adjoining valleys of *Gilgit* the same travellers got as far W. as E. long. 74° 25′, and as far N. as N. lat. 36° 20′, at the latter point reaching a glacier, perhaps of the Turkish watershed; but this district I suppose to lie without the limits of Tibet.

Our knowledge of the breadth of this region is not so precise; for the Turkish watershed has been actually crested at one point only, as already mentioned, viz., by Dr. Thompson, at the *Karakorum*, in 1848. A straight line drawn from that point to the south-westward, and perpendicular to the longer axis, meets the watershed of *Kashmir*, near the *Zoji La*, and measures more than 143 miles. In my explorations farther E., I reached a point—viz., the E. head of the *Chang-Chenmo* valley—more than 150 miles in the same direction, from the watershed between *sPiti* and *Rupshu*, but without attaining any knowledge of a Turkish watershed. Still E. of this our explorations are narrowed again by the diplomatic cordon of the Chinese, till they end at the points first mentioned, in the S.E. corner of *Nari-Khorsum*, leaving all the north-eastern quarter of that province comparatively unknown: but native reports attest the existence of a large belt of high pastoral uplands, thinly inhabited by Tibetan shepherds, and subject to the government of *Gar*. In Balti, the explorations above mentioned make the probable position of the Turkish watershed only about 100 miles (of direct map distance S.W. and N.E.) from the N.W. extremity of the Indian watershed, between *Hasora* and the *Kishenganga*. On the whole, we may consider the transverse breadth of *West-Nari*, from the Indian to the Turkish watershed, to average 150 miles, being less at its W. end in *Balti*, and more at its E. in *Nari-Khorsum*.

The total area given by these dimensions is 90,000 square miles; but it is impossible to say whether this includes any part of *Bong-Madma*—itself a very extensive district—or excludes the whole of it.

There is no marked natural boundary between *Nari-Khorsum* and *Maryul;* with the exception of one small part, where the present political or diplomatic boundary coincides with a transverse mountain range and a watershed of the *Sutluj* and *Indus*. This boundary approximates to the meridian of 79°; but the S. end of *Nari-Khorsum* projects half a degree further W., and the centre and N. of *Maryul* as much to the E., so that the mean line runs about S.S.W. and N.N.E. It roughly bisects the length of *West-Nari;* but if *Nari-Khorsum* be broader than *Maryul* (as seems probable), the former will contain about 50,000, and the latter 40,000, of the 90,000 square miles above assigned to the joint area.

A natural difference does exist in the general characters of *Nari-Khorsum* and *Maryul*, like that which I supposed to demark *Mangyul* from *U-Tsang;* viz., the greater depression of the valleys in *Maryul* as they approach towards the exit of their drainage by the gorge of the *Indus*, and consequently the greater warmth of their climate and suitability to agriculture; whereas the high and cold uplands, fit for little but pastoral uses, occupy the greater part of *Nari-Khorsum:* but the two characters are somewhat blended in parts, nor separated by any geographical landmarks; and the line of division lies rather to the W. of the diplomatic boundary above noticed, so as to include the eastern parts of

Maryul with *Nari-Khorsum.* More explicit information on this head will follow.

Geographical Subdivisions.—Before describing the geographical details of these countries, I shall state the names and general positions of their principal subdivisions. *mNaris*, commonly *Nari*, signifies the *Clear* or *Pure*, an appellation appropriate enough to the fine air and water of this country. *Khorsum* signifies the *Three Continents*, or circumscribed tracts, into which Tibetan geographers have divided the province: these are—*Ruduk* on the N., *Guge* on the S W., and *Purang* on the S.E. *Ruduk* is said to be encircled by lakes, *Guge* by rocks, and *Purang* by glaciers or snowy mountains; and though somewhat open to criticism in detail, these expressions have a general significance founded on facts of the natural geography. *Ruduk* is traversed, though not encircled, by the largest of all the lakes of *Nari*, and contains others of less note; and the prevailing character of the whole district is a succession of flat lacustrine valleys, similar to those that contain the existing lakes, and often perhaps exhibiting marks of others now extinct, such as the Tibetan peasant can appreciate in the gross as well as the European geographer. If we take the Tibetan word *brak* (v. *tak*) in the same sense as the geologist uses our word *rock*, and with its own particular import of precipitous form, we shall find it fully characteristic of the central parts of *Guge*, where the precipices of conglomerated alluvial earth are unequalled in all *Nari*, and probably in the whole world. In like manner the mountains of *Purang* are actually pre-eminent for snow—a joint consequence of their great elevation, southerly position, and close connexion with the Indian Himalaya.

The *Three Continents* above described do not exactly express the existing political subdivisions of *Nari;* at the present day they add a fourth district, *Gar*, which occupies the central part, and is dignified by the seat of provincial government bearing the same name, which signifies a *Cantonment;* and *Gar* is, in fact, a large encampment of tents, and not a town of houses: *Garoo, Gartokh, Gardok, Gortope*, &c., are Indian and English corruptions of this name. *Namru* is a small subdivision, on the west side of *Gar* proper. *Tashikang*, i. e., *the August Abode*, a monastery giving its name to the district upon the Indus next to the *Ladak* frontier. *Seng*, i. e., *the Lion*, subdivided into *Seng Tot* and *Seng Mat*, i. e., *the Upper* and *the Lower Lion*, is a large district at the head of the Indus, and either gives or takes the name of that river. *Matsa* lies in the unexplored part of *Nari*, at its N.E. end.

Purang is subdivided into four districts; viz., *Purang* proper, the seat of the Tibetan *dZong*, or district government, in the S.W.;

Kangri, i. e., the *Iceberg* or *Glacier*, on the N.W., containing the lake (*Manasarowar*) and mountain (*Kailash*) that form the *Tirthapuri*, i. e. *Fulfilment of Indian Pilgrimage; Horba*, i. e. *Turkey*, extending up to *Maryum La* on the S.E.; and *Bongba*, or *Little Bong*, a political or geographical connexion of the great *Bong-Madma*, on the N.E.

Guge was formerly called *Zhangzhung*, but the latter name is now obsolete. It contains at present two of the Tibetan *dZong*, one enthroned at *Tsaprang* on the *Sutluj*, in *Guge* Central and Proper, and the other at *Daba* in the south-eastern quarter. *Rongchung*, i. e. the *Little Valley* district, lies to the W., on the lower part of the *Sutluj*, and *Chumurti* to the N.W., next *Ladak; Tsotso*, on the extreme W., though now-a-days a political appendage of *Tashikang* and *Chumurti*, under the government of *Gar*, belongs in physical geography rather to *Maryul*.

Ruduk comprises—*Ruduk* Proper, the seat of a *dZong*, upon the *Ladak* frontier, south of the great lake of *Pangong; No* to the N. of it; and *Tsaka*, i. e. *the Salt-field*, an extensive but unexplored tract to the eastward. There may be other subdivisions of this district, but little is known about it.

Maryul signifies in Tibetan *the Low Country*, a term appropriate to the character of its inhabited valleys as contrasted with *Nari-Khorsum;* but the name is now obsolete, and superseded by the modern appellations of *Ladak* and *Balti*. *Ladak* occupies the larger and south-western portion of the country, which is inhabited mostly by Budhists; and Balti the extreme N.W., of which the population is exclusively Musalman; but there is no definite geographical boundary between the two, and the political division is somewhat complicated. The boundary between the Musalman and Budhist Tibetans crosses *Maryul* about the meridian of 76¼° at its S. end, and 77° at the N., leaving perhaps one-third of the whole area, or 13,000 square miles, to the former, and two-thirds of it, or 27,000, to the latter; but some of the Musalman districts were disputed, and others always possessed, by the Budhist Principality, up to the time of the Dogra conquest (1835-42), and the latest boundary of the *Gyalfu* (i. e. *King*) of *Ladak* reached, at its S.E. end, to the meridian of 75¾°, including about 2,500 square miles of the Musalman country in that quarter; so that *Balti* is thus reduced to little more than 10,000 square miles, and *Ladak* extended to near 30,000.

The name of *Ladak* has a local as well as a general sense; belonging originally, and still in particular, to the central district in and about the valley of the *Indus*, and in the middle of which is situated the capital *Le* The S.E. quarter of this is called *Rong*, which is properly a mere geographical term, denoting a *deep Valley*. *Hanle* occupies the south-eastern extremity of *Ladak*, next to *Nari-Khorsum;* and *Kakzhung*, on the highest part of the *Ladak Indus*, is geographically the northernmost part of this district; though the Tibetans make it politically an appendage of *Rupshu*, which is a large district in the S. *Tanktse*, in the N.E., includes *Pangong*, so called from the lake of that name, adjoining *Ruduk* of *Nari*. *Nubra* is the northwestern district, and *Yarma*, i. e. *Upper Nubra*, a northern subdivision of it. E. of this there is a large tract of savage, uninhabited country, belonging politically to *Shayok* of *Tanktse*, but without any general name of its own, and for want of which it may be called after that place.

Zangskar, i. e. *White Copper*, though politically a province of itself, is geographically part of *Ladak*, in the S.W. quarter. It includes *sTot*, i. e. *Upper*, and *Sham*, i. e. *Lower*, *Zangskar*; *Lungnak*, the *Black Vale*, in the S.E.; and *Kharnak*, the *Black Castle*, which is geographically the N.E. corner of *Zangskar*, though at present politically attached to *Ladak* proper, which adjoins it on the N.

Purik comprises roughly all the Musalman country at the western extremity of *Ladak*, towards *Kashmir* and *Balti*: its chief districts are *Purik* proper, *Suru* or *Suru Kartse*, and *Hembaps*, which last the Kahsmiris improperly call *Dras*, after its chief village.

Besides those above mentioned, there is the district of *sPiti*, including its southern subdivision *sPin*, which belonged politically to *Ladak*, till it was annexed to the hill provinces of British India; and in most of its geographical characters it assimilates with the former country, though in position it projects further to the S.W. than the general run of the Tibetan frontier, and alignes rather with the valleys of the Indian *Himalaya*; the Indian watershed, however, not being easily definable in this quarter.* The area of *sPiti* is about 1,500 square miles, lying without the limits above assigned to *Nari-Maryul*, which, if added to *Ladak*, will make the total area of the *Gyalfu's* Principality, as it existed in 1835, fully 30,000 square miles (at which it was estimated by Moorcroft in 1821).

The districts of *Balti* are, I believe, mostly named after their *Khar*, i. e. *Castles*, or seats of local government; but the central district, containing the capital town *sKardo*, is probably *Balti* Proper. The others are—*Chorbat Khapalu* and *Kiris*, in the northern valley, on the E.; *Khartaksho Tolti* and *Parkuta*, on the main *Indus* to the S.E.; *Shigar*, with the subdivisions of *Barsho* and *Braldo*, on the N.W.; *Rongyul* or *Rongdo*, in the valley of the Lower *Indus*, on the W.; and *Hasora* at the westernmost extremity of *Balti*, and of all Tibet. As I have seen nothing of *Balti* myself beyond the small corner of *Chorbat*, and my knowledge of the rest is secondhand, what little I say of it must be taken "cum grano."

I subjoin a classified list of all the Tibetan provinces and districts mentioned in the preceding pages.

- Bod, *or* Bodyul
  - Kham, Khamyul, *or* Bodchen.
  - Bod, *or* U-Tsang
    - Takpo.
    - Kongbo.
    - dvUs, v. U.
    - Tsang.
  - Lho-Pato, *or* Lho-Bruk, &c.
  - Nari
    - Mangyul
      - Zangzang.
      - Kyado.
      - Semukul.
      - Saka.
      - Nyugu.
      - Tradum.
      - Troshot.
      - Shamtsang.
      - Tingri.
      - Shelkhar.
      - Nyanam.
      - Khyirong.
      - Chamshen-Tsuglakang.
      - Kungtang.
      - Bong-Madma.
    - Nari-Khorsum
      - Purang
        - Bongba.
        - Horba.
        - Kangri.
        - Purang.
      - Gar. .
        - Seng
          - S. Tot.
          - S. Mat.
        - Gar. Namru.
        - Tashikang.
      - Matsa.
      - Ruduk
        - Tsaka.
        - Ruduk.
        - No.
      - Guge .
        - Daba.
        - Guge. Tsaprang.
        - Rongchung.
        - Chumurti.
      - Tsotso.
    - Maryul . . .
      - Ladak
        - Hanle.
          - Kakzhung.
        - Rupshu.
        - Rong.
        - Tanktse. Pangong.
          - Shayok.
        - Nubra. Yarma N.
        - Ladak. Le.
          - Kharnak.
        - Zangskar
          - Zangskar
            - Sham-Z.
            - sTot-Z.
          - Lungnak.
        - sPiti. sPin.
        - Purik
          - Purik.
          - Suru. Kartse.
          - Hembaps.
      - Balti .
        - Chorbat.
        - Khapalu.
        - Kiris.
        - Khartaksho.
        - Tolti.
        - Parkuta.
        - Balti. sKardo.
        - Shigar
          - Barsho.
          - Braldo.
        - Rongyul *or* Rongdo.
        - Hasora.

The names above given are *proper names*, of fanciful or trivial meaning when they have any. But the Tibetan geographers have also some purely natural divisions of their country. They divide the breadth of all *Nari* and *U-Tsang* into three zones, each extending along the whole length, viz., the *Lhogyut*, i. e. *South-lands*, which occupy the third next to the *Indian Himalaya*, communicating by passes into *Monyul*, and sometimes penetrated by the heads of Indian rivers; the *Zhunglam*, i. e. the *Mid-way*, or central zone, which is traversed longitudinally both by the main axis of drainage and by the principal line of communication between the several provinces of W. Tibet; and the *Changyut*, i. e. *North-lands*, which occupy the northern zone towards *Turkistan* and the *Rundur* country.

Each of the chief provinces has its share of all three zones; reckoning from N.W. to S.E., the *Changyut* will comprise *Shigar*, *Kiris*, *Khapalu*, and *Chorbat*, of *Balti*; *Nubra*, *Shayok*, *Tanktse*, and *Pangong* of *Ladak*; *Ruduk*, *Tsaka*, and *Matsa*, of *Nari-Khorsum*; and *Bong-Madma* of *Mangyul*: the *Zhunglam-Rongyul*, *Central Balti*, *Parkuta*, *Tolti*, and *Khartaksho*, of *Balti*; *Purik Proper*, *Ladak Proper*, *Rong Ladak*, and *Kakzhung*, of *Ladak*; *Tashikang*, *Namru*, *Gar Proper*, and *Seng*, of *Nari-Khorsum*; and *Shamtsang*, *Troshot*, *Tradum*, *Nyugu*, *Saka*, *Semukul*, *Kyado*, and *Zangzang*, of *Mangyul*: and the *Lhogyut*—*Hasora* of *Balti*; *Hembaps*, *Suru*, *Zangskar*, *Lungnak*, *Rupshu*, *sPiti*, and *Hanle*, of *Ladak*; *Tsotso*, *Chumurti*, *Rongchung*, *Guge*, *Daba*, *Kangri*, *Purang*, *Horba*, and *Bongba*, of *Nari-Khorsum*; and *Kungtang*, *Chamshen-Tsuglakang*, *Khyirong*, *Nyanam*, *Shelkhar*, and *Tingri*, of *Mangyul*.

The chief natural significance of the three Tibetan zones consists in the arrangement of the rivers, which generally preserve a main trunk in the *Zhunglam* with branches in the *Changyut* and *Lhogyut*; and in the differences of hygrometric climate, which make the southern zone more moist and snowy than the others. But the classification appears to be disturbed here and there, either by natural irregularities in the plan of the country, or by the intrusion of political elements: nor can I answer for its application farther E. than *U-Tsang*. Thus the *Zhunglam* is by some considered to pass from *Gar* into *Kangri* and *Horba*, because the *Gyalam*, or artificial *High road*, here lies to the S. of the natural *Mid-way*; and *Demojong* and *Lhopato* may be reckoned parts of the *Lhogyut*, though anomalous projections beyond the natural limits of the zone.

Another and important natural classification will be described further on.

Vertical Formation of the Table-Land.—Having sketched the general plan of *Nari* in horizontal extension, I must now explain its vertical relief, as this is developed on such a gigantic scale that no true idea of the country could be formed without equal attention to *the third co-ordinate.*

The area of *West Nari* consists then of a dense mass of huge rocky mountains, whose habitable, or even accessible valleys, bear but a small proportion (with one remarkable exception) to the solid mass of mountain, too steep and high for any human uses. The loftier summits rise in all quarters to an elevation of 4 miles above the sea-level, a few of the highest (as yet measured) reaching nearly to 5; and the deepest of all the fissures at its very lowest point (the exit of the *Indus* at *Acho*) is still nearly 1 mile above the sea, whilst there is only one other (the exit of the *Sutluj* at *Namgya*) that descends to less than 2; so that the mean elevation of the whole is probably near 3 miles.

For a clearer comprehension of this mass it may be divided into its two component parts, viz. solid rock *in situ*, which constitutes all the mountains, and by far the greater part of the whole, and a much smaller proportion of alluvial soil composed of the débris of those rocks, and laid out in horizontal deposits which, accumulated in the fissures of the mountains, form the bottoms of the principal valleys.

The alluvial beds, as now existing, appear to be only the ruins of a former and much greater structure: entire plateaux of any considerable extent are the exceptions to the rule, and in most places they have been broken down again by some destructive agency, leaving nothing but a collection of huge fragments lodged in the hollows of the mountains most favourable to their retention, so that the aggregate area of all the remnants can hardly exceed, I suppose, a tenth part of the whole country. In some instances the destruction has been so great as to remove the old alluvium down to the very bottom of the mountain fissures, and a main valley even is thus degraded into a gigantic ravine, with its river running through a bottom almost impassable in a sharp angle of bare rock, whilst remnants of the ancient deposits still adhering to the mountain flanks thousands of feet above, attest a former valley expanded to the breadth of a mile or two.

The alluvium, in what may be presumed to be its original completeness, appears to attain its maximum elevation about 16,200 feet: at 16,500 it may be seen just rounding off the mountain feet and connecting them by undulated slopes with the lower flats, but at 16,000 it is often thoroughly developed in horizontal plateaux; and the uniformity of the maximum limit of elevation in all parts of the country is as remarkable as the invariable horizontality of the strata. The ancient plateaux are generally in such a fragmentary state that it is difficult to assign a minimum elevation for their original surface, but I doubt whether this is to be found below 14,000 feet; and supposing it to have been laid out everywhere—as the very wide distribution of the fragments attests—at an average height of 15,000 feet, the old alluvium must have occupied nearly half the area of the whole country, often expanded into wide plains and broad connected valleys, whilst the mountains were more separated, isolated, and comparatively lower than we now find them.

The destruction of the primary alluvium has been succeeded or accompanied by the formation of a secondary order of beds, which abound in the lower valleys (where the primary banks have been most destroyed)—either simply as flat bottoms, or as high scarped terraces or broad shelving slopes between the mountain foot and river bed, or as "heaps of shot rubbish" spreading out of the mouths of lateral ravines into the main valleys; and most of these again exhibit marks of extensive subsequent destruction similar to that of the primary banks, though on a smaller scale: they are remarkable also for constituting the sole or chief sites for human habitation and agriculture.

The same cause has created a secondary class of valleys or ravines in the masses of the primary alluvium, where the original structure has survived wholesale ruin: these are oftenest mere ravines, with steeply scarped flanks, though measured by hundreds of feet, perpendicular cliffs of 100 vertical feet being common in *Ladak;* but in *Guge,* where the alluvial deposits are developed on the largest scale, the ravines become expanded into respectable valleys thousands of feet deep, with huge mountainous sides, sometimes in steep smooth slopes like the earthwork of an embankment, or broken into cliffs that vie with the grandest precipices of rock, and surpass them in the extraordinary artificiality of their appearance.

Irregularities in the original composition of the alluvium, and in its subsequent destruction, have caused a number of curious appearances in the secondary ravines. The cliffs are sometimes full of holes, such that the Turks have called one of them on the *Yarkend* road *Kuptar-Khana,* i. e. *Pigeon-House*—and the pigeon holes occasionally expand into caves large enough for the abode of Tibetan hermits. Sometimes alluvial aiguilles rise in sharp isolated points from banks otherwise crumbled down into gravelly slopes, and occasionally the crags and pinnacles of the more precipitous banks are so fantastic that it is difficult to persuade oneself they have not been cut into shape by the hands of man, or seen at a distance that they are not houses and castles; whilst some of the largest ravines —as those of central *Guge*—present the appearance of whole streets and cities of the most gigantic proportions and wonderful architecture.

Sometimes rock *in situ* protrudes through the alluvial sides of the secondary ravines, and changes their appearance to that of a primary valley between mountains, or projects above the upper surface so as to form isolated hills in the midst of the elevated plateau; and less frequently an alluvial mass itself is so wrought as to resemble closely one of the smoother hills of solid rock.

The composition of the alluvial beds is well exhibited in the vertical sections so frequently cut through them. They consist of the débris of the surrounding mountain rocks disposed in horizontal strata, often very strongly marked by alternations of large boulders, coarse gravel, and fine sand or clay, the individual fragments of which are generally more or less rounded. Limestone seems to impart a great consistency to the alluvium, and deposits of this material sometimes occur, which are aggregated into a solid rock too hard to be broken by the hand, though showing no signs of geological metamorphism. The deposits of *Guge*, which are the most extensive we know of, appear to consist of more finely comminuted material in their central part, where furthest distant from the mountains, the great ravines there being flanked by cliffs half a mile high, which exhibit throughout their thickness a fine homogeneous clay with little gravel in it. The stratification of the alluvium seems to be horizontal in all cases, or at most very slightly sloped from the mountain foot to the valley middle, in accordance with the existing directions of the drainage. I have once or twice seen small faults where some of the strata had become canted from the horizontal for a few yards, but never anything like a general disturbance of the original position.

It is in the larger alluvial deposits of *Guge* that the fossil remains of large mammalia are found, which have reached the hands of Indian geologists under the *Bhotiya* name of *Bijli-Har*—a translation of the Tibetan *Duhrus*, i. e. *Dragon bones*, which the savans of *Kumaon* have translated into *Lightning-bones*. I never heard of these in any other part of *Nari*, nor has any English traveller actually found their site, but they no doubt come from the ravines in the vicinity of the *Sutluj*. The small shells which occur in the superficial strata of alluvium, close to the margins of some of the lakes, appear to be comparatively recent, and some of them perhaps of species still living in the lake water.

Being no geologist I cannot undertake to give a description of the rocks which constitute the mountain masses; but a few facts are obvious to ordinary observation. Granite and Metamorphic Schists are decidedly the most prevalent rocks throughout *West Nari*: Greenstone and Porphyries are also common, and Serpentine and Statuary Marble occasional: Limestones are abundant, particularly in the southern zone, and often contain vast quantities of Coral and other Fossils: Fossiliferous strata, containing Ammonites and Belemnites, have been found at several points upon the Indian watershed all the way from *sPiti* to the *Saligrami Gandak* of *Naipal*—a line of four or five hundred miles.

Among the metamorphic rocks there are some remarkable specimens of Conglomerate to be seen in perfection on the left bank of the *Indus* in central *Ladak*. These consist of ancient alluvial masses precisely similar to the existing ones, converted bodily to a sub-crystalline rock without any partial disturbance of the strata, but canted to a dip of 30° or 40°, in masses which appear (from the town of *Le*) to contain a continuous succession of strata 2 miles thick. The alternations of large boulders, coarse gravel, and fine sand, are all represented in bands of solid rock, projecting or receding from the average surface according to their greater or less compactness and resistance to wear; and all the boulders and pebbles of the original detritus may be seen embedded in their sandy matrix, but so thoroughly cemented together by the metamorphic agency, that they break through with the fracture of the mass sooner than separate from it.

Though unable to discuss the geology of the country in detail, I think it as well to state the general impression which I have derived from a long and extended observation of its surface. The immense extent of the existing alluvium, and the uniformity of its maximum elevation, lead me to infer that it must have been deposited under a general sea covering the whole country (in which it so occurs), and not by lakes, much less by rivers. The existing lakes are comparatively mere puddles, made by the drainage of the higher and smaller mountain ravines stagnating in slight hollows of the old alluvium; and the gigantic ravines in the ancient plateaux are out of all proportion to the contemptible rivulets that now creep along their bottoms, wandering over beds a thousand times too large for them. The universal horizontality of the strata, and the conformity of all the alluvial ravines to the existing directions of drainage, whilst the existing drainage itself seems quite inadequate to their formation, show that they must have been laid out whilst the framework of mountain was in its present form, and that no material convulsion of the rock could have accompanied the general upheavement; and the elevation which the alluvium attains in excess of the gorges by which the present drainage has an exit to the lower regions proves that it cannot have been deposited in the existing elevation of the table-land above the general ocean, but at the bottom of that ocean in a former state of the earth, and upheaved afterwards by the equable raising of a whole continent.

I must observe here, that though I have myself divided the alluvial beds into two classes, primary and secondary, this is merely for facility and brevity of description; and a much more complicated and extensive classification may be necessary for a true account of their geological formation.

The preceding account of the rocks and alluvial plateaux, as met with in *West-Nari*, however geologically imperfect, will enable the reader to understand the natural classification of *Rong* and *Changtang*, into which the Tibetans divide the whole of their table-land; and the annexed figure will assist the explanation.

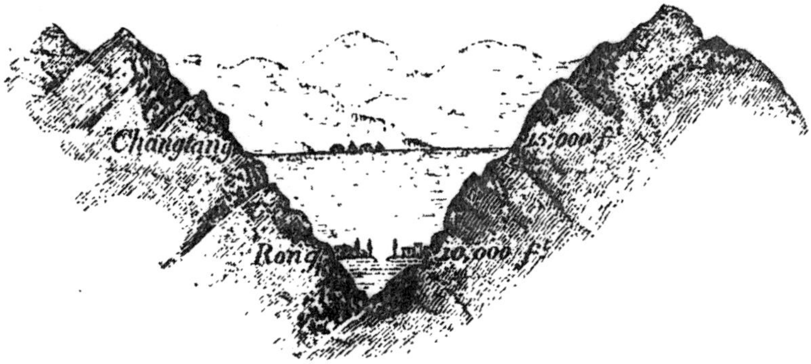

Rong signifies a *deep Valley*, low and warm enough for agricultural occupation, and generally a country containing such valleys. *Changtang* literally means the *North Plain*, but in common parlance an elevated plain, or wide open valley, too high and cold for any but pastoral uses, and generally a country of such valleys. The former description of country is produced by the destruction (or at least the absence) of the alluvial deposits, which in lowering the valley bottom also contracts it, and steepens its sides of mountain rock: the latter results from the preservation (or presence) of the alluvium in its original plateaux at great elevations, where the mountain masses occupy less of the area, diminished also in relative height and steepness.

Changtang accordingly is the prevailing character of *Nari* in its E. and central part, or *Mangyul* and *Nari-Khorsum*, where the beds of drainage are still very elevated; and *Rong* at its W. end, or *Maryul*, where the bottoms descend towards the main exit of the drainage by the gorge of the *Indus*. *Mangyul* is said to be *Changtang* in all its northern and central zones, and *Rong* only where some of the *Naipal* valley-heads penetrate its southern border. *Nari-Khorsum*, though mostly *Changtang*, is not without its *Rong*: in *Purang* the *Mapcha* river makes its exit by a valley very similar to one of Middle *Ladak*; and *Guge* is a strange compound of *Changtang* and *Rong*, as its secondary or alluvial valleys afford ample sites for agricultural villages in their lower parts, though their upper ends enter the pastoral uplands at the top of the great plateau: *Tsotso*, the exit of the *Rupshu Sutluj*, is also *Rong*. *Maryul* has its share of *Changtang*, but chiefly at the E. end of *Ladak*; and all *Balti*, I believe, is exclusively *Rong*, unless the tract which some call "the *Steppe of Deosu*" be a little isolated scrap of *Changtang*. The western limit of *Changtang* crosses *Ladak* about the meridian of 77½° S. of the *Indus*. and 78½° N. of it; but the *Rong* of Upper *Ladak* projects to E. of the general line in the central part, and at the S.W. end there is a small detached piece of *Changtang* so far as 76½° (viz., *Rangdum*, in the high ground that divides the heads of the *sTot-Zangskar* and *Suru* valleys). The districts of *Changtang* in *Ladak* are—*Pangong*, with *Khargyam*, of *Tanktse*; *Hanle*, with *Kakzhung*; *Rupshu*, with *Lingti*; *Kharnak*; and *Rangdum*, above mentioned: and the joint area of these is about 10,000 square miles, leaving for the *Rong* districts of *Ladak* 20,000; but both of these include a considerable proportion of valleys utterly unoccupied either for agricultural or pastoral uses, and scarcely habitable from extreme narrowness and sterility.

In elevation the lowest limit of *Changtang* may be reckoned about 13,800 feet, and the highest of *Rong* 14,300, the mean line of demarcation being 14,000 feet.

The term *Changtang*—*Highland* or *Upland*—must not be confounded with the *Changyut*, which signifies *Norland*, as the former is essentially independent of the three zones, and may exist indifferently in any of them. though it no doubt originated in the prevailing character of the north country. The term *Changpa* strictly signifies a *Norman* or *Norlander*, an inhabitant of the *Changyut* in fact; but in actual parlance it is often applied to the inhabitants of the *Changtang*, denoting merely a *Highlander* or *Uplander*, and commonly a shepherd of whatever region. *Rungba* is the inhabitant of the *Rong* country, a *Lowlander*, and generally an agriculturist.

Mountains.—The mountains that compose the bulk of *West-Nari* are not easily understood or defined. On ascending the highest passes, we can seldom see anything but a contracted view of mountain tops on all sides, looking very like chaos: no general view of the range under our feet is ever obtainable, as the passes naturally select the ravine-heads and lowest points of the ridge, which are not only flanked, but often almost surrounded, by the higher summits; and the valleys are commonly so steep and narrow, especially in the *Rong* country, that the view can hardly ever penetrate to an alluvial bottom, and the sight of any inhabited place from a pass top is most unusual. When travelling along the bottoms of the valleys, we generally see nothing but a narrow tortuous passage, between steep rocky walls, shutting out all extended view, and rather concealing than exhibiting the mountain ranges of which they form but the mere lowest outworks: consequently it is only by an extended series of observations and inferences, joined and assisted by maps, that any regular arrangement of these mountains can be distinctly established; and my account of them is liable to error in proportion to the defects of my own map.

The general plan of the mountain system appears to me to consist of a series of parallel ranges running right across the breadth of the table-land in a direction so extremely oblique to the general extension of the whole as often to confound the one with the other, or to convert the transverse direction to a longitudinal one. The annexed figure may help to explain this.

Short transverse necks connecting the main ranges in some parts, and cross fissures cutting through them in others, together with projecting spurs of a secondary order, will suffice to convert the supposed primary arrangement into all the existing varieties of valley and drainage. Such connecting necks, when above 18,000 feet, become more or less confounded with the main ranges, and if not above 17,000 feet, often appear as low watersheds, just dividing the heads of two valleys lying in one line but draining opposite ways. Secondary spurs also may be so high and so obliquely joined to the primary ranges as to make it difficult to distinguish between the two; and the cross fissures may sometimes admit a main river to pass through a main mountain mass, in which case the continuity of the range is often evidenced by the extreme narrowness of the rocky gorge, or height steepness and geological correspondence of its sides.

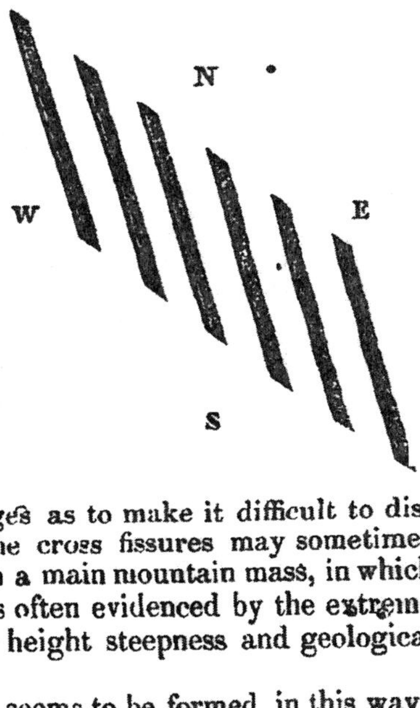

Much of the Indian watershed seems to be formed in this way; the great snowy peaks lying mostly on the terminal butt-ends of the primary ranges, sometimes widened by lateral spurs; and the Tibetan passes crossing the low connecting links, whose alignement forms the main watershed, but not the main mountain-crest. From what little I know of the Turkish watershed, I should suppose the same formation to exist there also.

The general extension of *West Nari* being nearly S.E. and N.W., I consider the component mountain ranges to run S.S.E. and N.N.W., but more generally and more certainly they lie within the third and sixth sextants of the compass; and in the few cases where they appear to enter the first and fourth, they seem capable of being resolved into the butt-ends of the others connected by inferior links.

The longest ascertained range in the interior of the country is the one flanking the right of the *Ladak Indus*, and dividing it from the valleys of *East Balti*, *Nubra*, and *Tanktse;* the length of this range is about 225 miles; its direction nearly S.E. and N.W.; and it runs almost all through *Ladak*, beginning rather short of the South-East frontier, but extending as much beyond its North-West into South-East *Balti*. The ranges on both sides of this chain are generally more meridional in their direction, so as to meet the left of the *Indus* and right of the *Nubra* river very obliquely; and they cannot be traced continuously to such a length as the preceding; the longest *ascertained* being rather more than 100 miles, dividing *Nubra* from *Kumdan*, but possibly extending further beyond the Turkish watershed. In *Nari-Khorsum* the longest ascertained range is that dividing *Guge Chumurti* and *Koyul* (of *Hanle*) from *Gar Namru Tashikang* and *Upper Kakzhung*, and in its middle part the *Indus* from the *Sutluj*, having a length of 150 miles, and direction of S.E. by S., N.W. by N. The range dividing *Horba Kangri* and *Gar* from *Seng* is perhaps longer, though our positive knowledge of it does not exceed 150 miles, approximately S.E. and N.W. The Peak of *Tise* or *Kailash* is the end of a long spur projecting from the S.W. face of this. There is also a range of more than 100 miles dividing *Ruduk* from *Tashikang* and *Kakzhung*, with a direction nearly S.E. and N.W.

I had no opportunities for measuring the height of inaccessible summits in *Ladak*, and could merely estimate some of the most conspicuous by view from lower barometric stations, in which estimates the most experienced mountaineer is liable to err by *thousands* of feet. On the whole I should suppose 20,000 feet to be a very common height for the average crest of the main ranges in most parts of *West Nari*. In the *Changtung* of *Ruduk Hanle* and *Rupshu* there is a good deal of mountain something lower, and some considerably lower in the *Rong* of *Purik*, and in most parts of *Balti*. The long range of *Central Ladak* does not average perhaps more than 19,000 feet; it is still less at its N.W. end, towards *Purik* and *Balti*, but at the S.E. contains some snowy summits probably much above that height. The range south of Central *Ladak* seems above the average height, and its most prominent peak, opposite to *Le*, is perhaps 22,000 feet. The mountains dividing *Tsotso* from *Hanle* and *Chumurti* are also above the ordinary height, and as Gerard estimated some of their peaks at 30,000, and *Porgyul*, measured 22,600, appears from *Guge* to be quite lost among them, the highest are not improbably 23,000 feet. The short range flanking the west of the *Pangong* Lake appeared to me higher than any other in *Ladak*, averaging I should suppose 22,000 feet, with peaks of 23,000, the summits being covered with a wide belt of névée whilst my own barometric station at 18,800 feet was entirely free from all snow.* Dr. Thompson estimated one of the peaks at the head of the *Kumdan* valley at 24,000 feet; and from the top of the *Karakorum Pass* (itself 18,000 feet) he saw a range of mountains, still N. of the Turkish watershed, very snowy in August; therefore in all probability above four miles high.

In *Nari-Khorsum* my brother has measured a few of the peaks on and near the Indian watershed, by purely geometrical operations which assure the results within one or two hundred feet. The highest of these is *Kamet* (i. e. *the Snowless*—name doubtful), situated on the watershed of *South Guge* and *East Chongsa*, a most conspicuous landmark from all the elevated parts of *North Guge* and *Chumurti*, and also visible from *Almora* on the Indian side, where, however, its appearance is so modest that, till 1849, it remained unnoticed and unmeasured, though but 250 feet lower than the King of the Western *Himalaya*, *Nanda-Devi*. The main peak of this mountain is 25,500 feet, and the points of its two chief spurs about 24,000. Next to this is *Gurla Mandhata* or *Momonagli* (name doubtful), forming the end of a lofty spur that runs from the Indian watershed up the N.E side of the gorge of *Purang*, till it suddenly terminates on attaining its greatest height immediately south of the lakes of *Kangri*: its highest summit is 25,200 feet, and other points of the same mass 22,200, 22,500, and 22,800 feet. The peak of *Tise*, or the Indian *Kailásh*, which is purely Tibetan though in the *Lhogyut*, is 22,000 feet; and some of the main range behind it is—to judge from its snow—quite as high, though less conspicuous from the S. and W. A peak of 20,500 towards the S.E. end of the range dividing *Gar* from *Guge*, though slightly prominent from this the lower end, appeared to be equalled if not exceeded by the average crest at the *N.W.* end of the same range, between *Chumurti* and *Tashikang*. The summits, of 18,000 feet and upwards, that project from the Indian watershed of *Daba*, or are detached from it in *Kangri*, look like mere hills compared with the neighbouring mountains, and might be subtracted without causing much difference in the general appearance of the country.

The passes, which are commonly the lowest part of the ridges, generally exceed 18,000 feet on the principal ranges in *Ladak* and *Nari-Khorsum*; if less than 16,500 they may, in the *Chang-tang*, be little better than Tibetan valleys; but in *Purik* and

* A pass of 17,000 feet, which connects the lower end of this range with the long range of *Ladak*, looks more like a valley than a mountain crest, so much is it exceeded by the main ridges on both its flanks.

Balti some of the main ranges perhaps may be crossed at such elevations. The highest pass I crossed in the course of my own travels was the *Zulang La*, leading from *Koyul* of *Hanle* into *Chumurti*, which measured (by boiling thermometer geometrically compared) 19,000 feet: but another pass of the same ridge distinctly visible a mile or two off seemed 500 feet higher, and this is the highest of which I have any cognizance. Passes between 18,000 and 19,000 feet are of the commonest occurrence all over *Ladak* and *Nari*, not excepting the second-rate range of *Middle Ladak;* and some of these are frequented by Turks and Tibetans even in midwinter.

The Tibetans call mountains *Ri*, and a mountain pass *La*. They have *proper names* for a few remarkable peaks and for all of the passes, but no general name for whole ranges; and when such appear upon our maps they are the misapplication of purely local names by English surveyors and European geographers. Thus the Tibetan *Para Lasa*, i. e. *Middle Pass*, the Turkish *Mus-Tag*, i. e. *Ice-Berg*, and *Kara-Korum*, i. e. *Black Gravel*, applied by the natives exclusively to the mere passes, and the Indian *Kailásh* to a mere peak, have been "raised to that wrong eminence" upon the map of Asia.

Valleys.—As the valleys of *West Nari* (in Tibetan *Lungba*) are the very essence of the habitable country, their names are generally identical with those of their inhabited places, which are often taken from the chief village, or from the *Khar*, i. e. *Castle*, or seat of local government; and the smallest ravines even have their names, in common with the hamlet pasture ground or encamping place contained in them, or names of their own when containing nothing. The terms *sTot*, *Yarma*, *Kongma* or *Kongnus* = *Upper; Parma* or *Parba* = *Middle;* and *Sham* or *Shamma*, *Mat* or *Madma*, *Yogma* or *Yognus* = Lower, are often used for the purpose of subdivision.

The general formation and arrangement of the valleys may be inferred in some measure from what has been said of the mountains that bound them, and the alluvial deposits that form their pavé. The prevailing direction of their main axes of drainage is of course like that of the mountain ranges, in the third and sixth sextants of the compass; but the exceptions to the rule are frequent, as a roomy valley may lie between secondary spurs of mountain, or even in the cross fissures of a main range, and will then usually have the opposite or transverse direction. The general arrangement may be briefly described to consist of one long central valley, traversing the whole *Zhunglam* from end to end, and a succession of shorter branches through the *Changyut* and *Lhogyut*, either joining the central valley or forming independent systems connected with the valleys of the *Indian Himalaya.*

The valley of the *Zhunglam* runs between 600 and 700 miles from S.E. to N.W., traversing the districts of *Seng-Tot*, *Seng-Mat* and *Tashikang* of *Nari*, *Kakzhung*, *Rong-Ladak*, *Central* and *Lower Ladak*, and *Khartaksho*, *Tolti*, *Parkuta*, *Central Balti* and *Rongyul*. Its minor branches on the N.E. side are very insignificant (so far as we know them); those upon the S. not penetrating to the *Lhogyut* include the *Kanji* valley of East *Purik*, *Yuru-Wanla* of *Lower Ladak*, *Koyul* of *Kakzhung*, and *Gar Proper*. In the *Changyut* the valley of *Shigar* is the only one of the secondary branches that joins the central valley direct.

All the rest of them, viz., *Khapalu*, *Yarma-Nubra*, *Kumdan* with *Chang-Chemno*, *lData* and *Chang-Parma* of *Pangong*, and *No* of *Ruduk*, with a few smaller branches from the S. side, first unite to form a great secondary trunk which runs the whole length of *Ruduk*, *Pangong*, *Tanktse*, *Nubra*, *Chorbat*, *Khapalu* and *Kiris*, parallel to the main valley of the *Zhunglam* (and separated from it by the long mountain-range of *Middle Ladak*) before joining it in Central *Balti*. E. of *Ruduk* the arrangement of the valleys is unknown.

In the *Lhogyut*, the small valley of *Hasora* joins the *Zhunglam* of *Lower Balti* direct, at the very W. end of all Tibet. Next to this is the system of *Purik*, which joins the central valley at the frontier of *Ladak* and *Balti* by a short transverse trunk uniting four main branches, viz., *Shingo* (name doubtful), *Hemhaps*, *Suru-Kartse*, and *Waka-Molbi* (or *Pashkyum*). The system of *Zangskar* is more extensive; its main trunk (which is very narrow and tortuous in its lower part) running almost perpendicularly up to the Indian watershed, with several large branches, viz., the *Zingchan* and *sTot-Zangskar* from N.W., and *Marka*, *North-Kharnak*, *Zangla* and *Lungnak* from S.E.; the last of these itself a large system, branching into *Khargya*, *Shade Shun*, *Lingti*, *South-Kharnak* and *North-Rupshu*. *South Rupshu* unites several small branches before joining *Tsotso*, and the latter enters *Hangrang* of *Upper Kanor* at the same point where *sPiti* joins it from the westward. The system of *Hanle* joins the Upper *Rong* of *Ladak* obliquely from the S.E., and is of much less extent than the others of the southern zone. The *Lhogyut* of *Nari-Khorsum* is formed into one long chain of valleys running S.E. and N.W., from *Maryum-La* to *South-Hangrang*, through *Horba Kangri* and *Guge*, expanded to a great breadth in *Guge*, and narrowed about *Kangri* by the encroachment of *Purang*, which is a separate valley descending to the *Indian Himalaya* of *Naipal*.

The valleys of *West Nari* may be made any length from 1 mile to 700, according to the degree of subdivision or generalization adopted by the geographer. The alluvial bottoms are much more continuous in the *Changtang* than in the *Rong* country. In all

Ladak the longest expanse of uninterrupted flat bottom is perhaps the valley of *Nubra*, where, from *Agam* to the foot of the *Yarma* glacier, we may travel for 80 miles in a straight line over alluvial flats without once crossing rock *in situ*, and with no other ascent or descent than the ups and downs of the secondary alluvial banks —which rarely exceed 100 vertical feet—or the general slope of the valley, which is only 20 feet in a mile and almost insensible; but not above a third of this length can be seen at once in a straight unbroken vista, and the views here are perhaps more extensive than in any other valley of the *Rong*. In the *Changtang* of *Ladak*, the longest extent of straight open valley is *Kakzhung* on the *Upper Indus*, where the same direction is preserved for about 50 miles without interruption of the alluvial flat, and but 200 or 300 feet of general slope, with a good breadth and unbroken view for two-thirds of that distance. This is probably equalled or exceeded by some of the valleys of *Nari;* besides *Guge*, which is altogether anomalous and will be separately described further on. Some of the best valleys of the *Changtang* moreover are occupied in part or whole by lakes, such as the *Tso-Moriri* of *Rupshu*, whose basin has a clear open length of 20 miles; and the *Pangong* of *Tanktse* and *Ruduk*, where the aggregate length of open valley is more than 100 miles, though broken into three or four vistas, so that not more than 25 or 30 miles can be seen at once. The lakes *Mapham* and *Langak* narrow without shortening the valley of *Kangri*.

The breadth of valleys is of necessity generally greater in the *Changtang* than in the *Rong;* yet expanded plains are the exception to the rule even in the *Changtang*, and the upland valleys very often assume the appearance of mere passages between walls of mountain. In most cases, and especially in the *Rong*, the alluvial bottoms are defined with great precision by the extreme abruptness with which the mountain sides rise out of them, in sheer rocky precipices or slopes inaccessibly steep and rugged; and very seldom is there any blending of the mountain and valley by inferior undulating hills.

In the *Changtang* of *Nari*, the valley of *Kangri* appears to attain an extreme breadth of 5 miles, besides lateral recesses partly occupied by the lakes: parts of *Gar*, *Tashikang*, *Ruduk*, and *No*, are also said to be expanded into small plains. In *Ladak* the valleys of *Chushul* in *Pangong*, *Kakzhung*, *Mid-Hanle*, and *Tega Leptra* and *Pangpok* of *South-Rupshu*, are often 3 miles wide, seldom less than 1, and sometimes above 4, and the greater part of this fair open plain.

In the *Rong* of *Ladak* the breadth of the valleys may be estimated to average no more than ½ a mile: ¼ mile is very common, and some of the bottoms occupied by fields and villages are even less than a furlong.

In some parts of the *Rong*, the gorge of mountain rock is so contracted as to be wholly occupied by the river, quite uninhabitable, and sometimes impassable. This occurs even in the main valley of the *Zhunglam*, both in Upper and Lower *Ladak*, which, though nowhere actually impassable, is often very narrow and rugged; and in the lower part of the gorge of *Zangskar*, much of which is totally impassable excepting by the ice of the river in winter. The lower part of the *Kharnak* valley can be traversed only by incessant fording and wading of the river in the dry season: and in the *Tilat* branch of the same valley, the road for some miles lies nearly as much in the river as out of it; at one point passing through a natural archway of rocks so low that a horseman wading the stream must stoop his head to get through, and at another through a rocky gorge which the Tibetans aptly call *Tarson-Migarson*, i. e. *Horse-way no saddle-way*, being nearly narrow enough to scrape the saddle off a horse's back, whilst the bottom is filled with dangerous accumulations of snow or water.

In many other valleys, such as that of the *Maryul Changyut*, both above *Nubra* in the *Shayok* and *Kumdan* quarters, and below it in *Chorbat* of *Balti*, though the bottom itself may be sufficiently wide and flat, the passage is obstructed by the depth of the river and its repeated contact with the foot of precipitous mountains; and in these cases the valley may be just passable in the dry season by frequent fording and wading, or at some points by climbing over the rocks of the mountain side (roads and bridges being in this country generally out of the question).

One of the widest expansions of valley in the *Rong* is in Central *Ladak*, where, for a length of 15 miles, the breadth varies from 4 to 8, exclusive of the further recesses of the side ravines; the mountain flank on the left (S.W.) being approximately straight, and that on the right (N.W.) broken into deep bays—the mouths of lateral ravines—which are separated by spurs of inferior but steep and rocky hills, narrowing to points as they advance and descend towards the *Indus*. The town of *Le* is situated in one of these recesses, 4 or 5 miles from the river, and it enjoys a wider southern aspect than almost any other place in the *Rong* of *Ladak*. The valleys of *Nubra* are seldom less than 1 mile wide, and in the lower part of *Yarma* as much as 3, with a still greater breadth at the junction of the two valleys about *Deskit*. Central *Zangskar* has an extreme breadth of 5 or 6 miles about *Padum* and *Karsha*, where the *sTot valley* unites with the mouth of *Lungnak* to form the valley of *Sham-Zangskar*. The greatest expansions are generally at the junctions of two or more valleys, but the irregular shape of such places often renders it difficult to define their breadth; and as all my statements on this head involve some guess-work, they might be modified by an exact survey of the ground with instruments.

If stratified alluvial deposit be considered essential to the formation of a valley, the greatest elevation of any valley bottom cannot exceed the uppermost limit of that alluvium, or 16,500 feet. The mountain hollows about this height are seldom more than mere ravines, though in one or two places they may be found looking very like valleys even up to 18,000 feet; and it is possible that some slight wash of alluvium has been deposited up to that height.

The lowest depressions are to be met with in the gorges by which the main rivers make their exit to India. The gorge of *Purang* is explored only down to *Taklakhar*, where the bed of the *Mapcha* river is still about 14,000 feet (doubtfully estimated from adjoining stations). The gorge of *Rongchung*, or Lower *Guge*, may be defined by drawing a straight line from Mount *Porgyul* to the high peaks on the south bank of the *Sutluj*, which will cut the gorge under the *Piming La*, where the height of the river bed may be estimated about 9,000 feet (being midway between *Shipki* 9,700, and *Namgya* 8,600): this place is narrow and rocky, I believe, with little remains of the alluvium. The gorge of *Tsotso*, discharging the river of *Rupshu* into *Hangrang*, is about 11,000 feet: the highest point which the Chinese government allow us to explore here is the *Takka rDozam*, where the river bed at 11,600 feet is extremely contracted between huge rocks, some of which have tumbled across the stream and formed a natural bridge more picturesque than convenient: but the conflux of the river with that of *sPiti*, where both debouch into the open valley of *Hangrang*, is (according to Gerard) only 10,200: and at both these points there are large alluvial banks much higher than the river bed. The gorge by which the *Indus* leaves *Balti* has been explored down to *Acho*, at the conflux of the *Hasora* river, which is nearly its lowest Tibetan point; and the height of the bottom here appears, from boiling observations by Mr. Winterbottom, to be about 4500 feet.

The valleys have, of course, every variety of elevation between the limits above assigned. All the main ones are remarkable for the gentle slope of their longer axis of drainage, which, indeed, is rarely sensible to the foot or eye, and completely lost in the ups and downs of the subordinate alluvial banks. The valley bottom of *Yarma-Nubra* is elevated 11,700 feet at the foot of the glacier that fills its upper end, close below the Turkish watershed, and 10,400 feet at its debouchure in *Mid-Nubra*, being a slope of 1,300 feet in a length of 55 or 60 miles, or a little more than 20 feet per mile. The Upper *Rong* of *Ladak* has a slope of about 1,500 feet in a length of 40 or 45 miles (between *Chumatang* 13,500, and *Shara* 12,000), or about 35 feet per mile; the slope of Lower *Ladak* is not so steep. Among the small lateral valleys—that of *Sakti* and *Chimra* has a slope of 1500 feet in 10 miles (between *Tagar* 13,000, and *Kalaktartar* 11,500), or 150 feet per mile: the valley of *Marka* is nearly the same: that of *Hanu* has a slope of 3,500 feet in 8 miles (between *Handumir* 13,000 feet, and *Nyubibrangsar* 9,500), or 440 feet per mile; and this is one of the steepest. In the *Changtang* the slopes are generally (but not always) much flatter: *Khargyam* of *Tanktse* has a fall of 2,000 feet in 20 miles (between *Long-Kongma* 16,000, and *Erat* 14,000), or 100 feet per mile. The valley of *Kakzhung* is so flat for 50 miles (between *Tedur* 14,000 feet, and *Loma-Sumdo* 13,800) that the slope could not be determined by my single barometric observations; and the 200 feet which I have assigned to it, or 4 feet per mile, is little better than conjecture. Probably much of the *Changtang* of *Nari* is as flat as this last, and in the lacustrine basins flatter: the *Tso Pangong*, for example, attests the perfect horizontality of its valley for a length of 100 miles.

Great Alluvial Plateau of Guge.—The valley of Guge is at once an exception to the general rule of formation, and the most remarkable of all the valleys of *West Nari* (so far as known to us). The usual dense mass of mountains here gives place to a large crater filled up with the alluvial deposits so as to make a great plain, which is elevated from 15,000 to 16,000 feet, and interrupted only by tremendous ravines that have been formed by the destruction of the original deposits along all the lines of drainage, or here and there by a few little islands of inferior hill.

The general shape of the valley is a very acute triangle, having its apex adjoining to the heads of *Purang* and *Kangri* on the S.E., and its base to *Hangrang Tsotso* and *Hanle* on the N.W. *Chumurti*, at the N.W. corner, is essentially part of the *Guge* Basin, being contained within the same main watershed, and distinguished only by the projection of inferior mountains above the level of the alluvial plain, which converts it to a hilly district, with narrow valleys like the ordinary Changtang.

The S.W. side of the triangle, bounded by the Indian *Himalaya* of *Kyunam* and *Chongsa*, runs from E.S.E. to W.N.W., and in this direction the continuous length of plain (broken only by the alluvial ravines) measures 120 miles, in a straight line from the S.E. end of *Gyanima* to the *Shelti* ravine between the *Kiobrang La* and *Hukyu*, which latter may be gathered from Gerard's account (though he is not himself explanatory on the subject), to be the limit of the alluvial plateau in that quarter. On the N.E. the length of continuous plain is 90 miles, measured from *Tsiring* on the *Chumurti* border to the *Sutluj* near *Kyunglung*, in a direction S.S.E. and N.N.W. along the base of the mountains that divide *Guge* from *Gar* and *Tashikang;* the length being here diminished by the mountains of *Chumurti* at the N.W. end, and a group of low hills dividing *Gyanima* from the *Sutluj* and *Kyunglung* at the S.E. The greatest transverse breadth measured S.W. and N.E. is about 60 miles, passing through *Rabgya-ling*, and its least breadth across *Gyanima* only 15 miles.

These dimensions give the alluvial plateau an area of 4,500 square miles; any small defect in which by encroachment of isolated hills is probably compensated by extensions of the alluvium into ravines of the mountainous enceinte.

The main axis of drainage, exhibited in the course of the *Langchen* river, runs nearly parallel to the S.E. boundary of the valley, leaving the greater breadth of plateau on its N. and the greater length on its S. side.

The elevation of the plain averages probably 15,500 feet. The highest parts of it at the foot of the mountains exceed 16,000 feet: from boiling thermometer measurements, made in the vicinity of *Tsiring*, I estimated the height there to be 16,200 feet; some of the highest terraces that come up to the foot of the *Niti* Pass are about the same (and the height well corroborated by my Brother's geometric measurements), and Gerard also gives 16,200 for the height of *Zingchan*, near *Bekhar*, which appears from his account to be alluvial ground. 15,000 feet seems to be the minimum elevation of the original surface where not corroded by ravines, and this only near the banks of the *Sutluj* where the plateau is least elevated.

The plains of Daba are perhaps generally less elevated than the other parts of Guge, by a few hundred feet, and especially about Gyanima at the S.E. extremity of all, where the plateau is connected with the N.W. head of the valley of *Purang*.

The general fall from margin to centre of the basin may be reckoned about 1,000 feet: but this is not uniformly distributed; its base varying perhaps from 10 to 40 miles, and the slope from 25 to 100 feet per mile.

Though clean flat plains often occur between the ravines for many square miles together, the general surface of the plateau is somewhat diversified by terraces of various level, usually separated by ravines, but sometimes rising one from the other in steps, of which formation the most remarkable instance that I met with is between *Ling* and *Mangnang*, where a direct breadth of four or five miles, between two of the main ravines, is broken into six different terraces, with a height of 2,000 feet between the top of the highest and the foot of the lowest, the separating steps being sometimes smooth steep slopes and sometimes broken cliffs.

A large proportion of the original plateau—perhaps half—has been destroyed by the ravines, which, conforming always to the present drainage, run in multitudes from the mountain-foot on both sides, uniting, deepening, and widening till they enter the great central ravine of the *Sutluj;* and the heads of some of them are remarkable for completely cutting off the alluvial deposit from the mountain wall, against which it must have originally rested, as at the foot of the *Niti* Pass, where the traveller to *Daba* has to descend abruptly 1,500 feet merely to rise again immediately 1,000, and in the *Shelti* ravine, between the *Kiobrang* Pass and *Bekhar* (described by Gerard). The ravines appear to moderate a little in the *Daba* quarter, and about *Gyanima* my Brother found there were none at all, the rivers running in shallow beds over the top of the plain.

I found the ravine of *Mangnang* at *Dakala* about half a mile wide across the flat bottom, and several others a quarter of a mile; and the great ravine of the *Sutluj* appeared to expand to the breadth of a mile between *Ling* and *Tsaprang*, though a little above *Ling* contracted to a very narrow gorge by rocks protruding through the alluvial sides. The dimensions from crest to crest of the ravine sides are commonly double and treble of the bottom breadths, and about *Ling* and *Tsaprang* the erosion extends irregularly over an immense breadth, difficult to define, but amounting I should suppose to 10 miles.

1,000 feet is a very common depth for the side ravines, even in their upper part near the mountains; and in their lower part, towards the *Sutluj*, they deepen to 2,000, 3,000, and perhaps more. I found the bed of the river at *Ling* to be elevated 12,400 feet, whilst the summit of the alluvial plateau was about 15,700 on the N. bank, and 15,200 on the S.; making a mean depth of 3,000 feet, all broken precipice of pure earth: but further W. the dimensions must be still greater, for Gerard made the bed of the *Sutluj* under *Bekhar* 10,800 feet, and the alluvial plateau of *Zingchan* 16,200, so that the ravine must there attain the depth of a full vertical mile (but whether all pure earth does not appear from his account).

Mean Elevation of the Table-Land.—Our knowledge of *West Nari* is hardly sufficient for the construction of complete or exact sections of the country, and any calculation of its mean elevation would be rendered very uncertain by the number of assumptions required to eke out the few actual measurements: but it seems to me that in *Guge* nature has by brute force effected some approximate solution of the problem in question. The mean elevation is in fact the height of an imaginary horizontal surface, leaving as much mountain above it as would just fill the valleys below it; and the formation of alluvial plateaux by the degradation of the mountains that contain them is a partial realization of the supposed process; which in *Nari* however has been stopped short of completion, as vast masses of mountain still remain above the highest of the alluvial flats; and this partial result is again partly undone by the subsequent destruction of so much of the original alluvium.

But in *Guge* I think the defect of the alluvium may be set

against the excess of the mountains, and the general level of the existing plateaux taken with great probability to indicate, approximately, the mean elevation of the whole tract as limited by the surrounding watershed. The Indian watershed is so far advanced to the N. as to exclude most of the great *Himalaya*, and the mountains of the north-eastern boundary are neither very high nor broad, reckoned from the elevated alluvial base of 16,200 feet, so that the whole mass of mountain within the watershed line may be conjectured with reason as no more than sufficient to fill the gigantic ravines that everywhere detract from the solid of the alluvium: if this view be correct the mean elevation of all *Guge* will be somewhere about 15,500 feet.

The mean of all *Nari-Khorsum* I should suppose to equal that of *Guge;* for though the valleys elsewhere are often less elevated, there is no where else (that we know of) such a deficiency of mountain. The mean of *Ladak* I think can hardly be less than 15,000 feet; considering that a third part of it is *Changtang*, as high as *Nari-Khorsum*, that the valleys of the Rong narrow as they deepen into mere ravines, and that all through the central meridian, in *Zangskar*, Central *Ladak* and *Nubra*, we find 11,000 feet valleys between 19,000 feet mountains. In *Balti* the elevations are considerably less, but the area also is smaller, and a mean height of 13,000 feet would only just compensate the excess of *Nari Khorsum* over *Ladak*.

On these grounds I venture to assign 15,000 feet, *provisionally* and *approximately*, for the mean elevation of all *Western Nari;* and in this estimate, as in all others, my aim has been to prefer understatement to exaggeration.

Rivers.—The Tibetans seldom have any name for their rivers; they are generally satisfied to call every large river *Tsangpo* v. *Sangspo*, and every small one *Tokpo*, or any stream indifferently *Chu* (i. e. *Water*); and if any proper name be appended to either of these terms, it belongs less to the stream itself than to the inhabited valley or accessible ravine which it traverses. Moreover the inhabitants of the principal valleys sometimes call their rivers by the name of the place next above themselves, from which the stream comes to them, rather than by the name of their own district, as the *Nubra Sangspo* in *Khapalu* and *Chorbat*, the *Shayok Sangspo* in *Nubra*, and the *Kumdan Chu* at *Shayok*. The people of *Balti* call the large rivers *Gyamtso* instead of *Tsangpo*, which is little better than an abuse of language, however, as the word means *Great Lake*, and is properly applied to a sea or the Ocean. They also call the *Indus* the *Share Gyamtso*, i. e. *Eastern River*, after the quarter from which it flows to them.

The mythology of the Tibetans, being in advance of their geography, has excepted one or two of the large rivers from the general want of names, but even here the term *Khabap*, i. e. *Vomitory Cataract* or *Fountain*, applies more to the mere source than the whole river: these exceptions are the *Senge Khabap*, i. e. *Lion Cataract*, the *Indus* of *Nari-Khorsum* and *Ladak*; *Langchen Khabap*, the *Elephant Cataract*, the *Sutluj* of *Kangri* and *Guge*; *Mapcha Khabap*, the *Peacock Cataract*, the *Karnali* of *Purang*; and *Tachok Khabap*, *i. e. Horse Cataract*, the river of *Nari-Mangyul* and *U-Tsang*, and ultimately the *Brahmaputra*: but on the whole these names are not commonly used by the unlearned, and I very seldom heard the term *Senge Chu*, *i. e. Lion Water*, applied to the *Indus* in Central *Ladak*. In the dialect of *Kanor* rivers are sometimes called *Ti*; and in this way the *Guge Sutluj* is named *Lingti*, the river of Ling, from the great monastery of *Tot-Ling*; and the *Tsotso River Parati*, from the shepherds' ground of *Para* in *Karak-Bargyok*. The term *Samandrang*, a corruption of the Hindi *Samundar*, i. e. *Ocean*, is applied to the united *Sutluj* of *Kanor* by a barbarism like the *Gyamtso* of *Balti*.

The general system of the rivers of necessity corresponds with that of the valleys. The longest trunk of the *Indus* drains the *Zhunglam* from end to end of the whole 600 miles. Its great northern branch drains all the *Changyut* of *Ladak*, and half that of *Balti*, for a length of about 200, the remaining 50 miles of *Balti* being drained by the independent affluent of *Shigar*. The *Changyut* of *Nari-Khorsum* is probably for the most part lacustrine, but at least 100 miles of it in the *Ruduk* quarter, comprising all the basin of the *Pangong* Lake, though not in active effluence, is connected by channels of dormant drainage with the system of the Northern *Indus* last mentioned. Of the remainder we are absolutely ignorant, and may suppose it connected either with the same system, or with that of the Upper *Lion* river of the *Zhunglam*, or even with the *Tachock* of East *Nari*, or with Turkish rivers of *Khotan*, or divided between several of these; but the scanty volume of the *Lion* river proves that there is no active effluence of any extent from the northward, and the gradual subsidence of the *Pangong* Lake that this can receive little from the eastward.

The drainage of the *Lhogyut* is divided between the southern branches of the *Indus* and the heads of the *Sutluj*; the former drain all the southern districts of *Balti* and most of *Ladak*; the latter a small part of *Ladak*, and all the *Lhogyut* of *Nari-Khorsum*, excepting the small section of *Purang*, which is a separate system drained by the Tibetan head of the *Naipal Karnali*, one of the chief sources of the *Gogra* of *Oudh* and *Behar*. In *Maryul* the drainage of the *Lhogyut* is not so united as that of the *Changyut*. South *Balti* is divided between the river of *Hasora*,

D 2

which runs to the *Indus* direct, and that of *Shingo* (name doubtful), which first joins the *Purik* river in *Ladak*. In South *Ladak* the main branches of the *Indus* are two, viz., the river of *Purik* and that of *Zangskar*, both a good deal subdivided, and the latter, which is much the larger of the two, notable as the main trunk of the *Indus* in respect of water volume, and connected at its E. end with the small lacustrine basin of North *Rupshu*. The branch of the *Sutluj* that drains South *Rupshu*, also connected with the small lacustrine basin of the *Tso Moriri*, after traversing *Tsotso* and receiving the smaller branch of *sPiti*, joins the main trunk of the *Guge* river in *Hangrang*, and there forms the *Sutluj* of *Kanor*. The basin of the *Langchen* is lacustrine in its upper part, containing the *Konkyu* Lake in *Horba* without active effluence, and *Tso Mapham* and *Langnak* in *Kangri* with a partial or intermittent one.

The following is a rough estimate of the areas of these basins and their principal subdivisions :—

	Square Miles.				
Indus of Nari-Khorsum & Upper Ladak	..	..	..	14,000?	
,, Zangskar and Lungnak . .	..	..	5500	6,000	
,, Lacustrine basin of North Rupshu, without effluence .	..	..	500		
,, Purik and Shingo	..	..	..	3,800	
,, Lower Ladak and Upper Balti	..	..	..	2,700	
,, Northern Indus of Khapalu, Nubra, Shayok, &c. . .	..	..	9000	15,500	
,, Basin of the Pangong Lake, without effluence . . .	..	..	6500?		
,, Lower Balti, with Shigar, and Hasora	..	..	..	5,000	
Total of Indus — Rivers	..	..	..	40,000	47,000
Total of Indus — Lakes, without effluence .	..	..	..	7,000	
Lacustrine basin of Horba, without effluence	..	..	2000?	12,500	
Ditto of Kangri, with effluence . .	..	..	2500		
Sutluj of Guge	..	..	8000		
South Rupshu and Tsotso	2300	2800	4300	4,500	
Lacustrine basin, inactive	500				
sPiti and sPin	..	1500			
Hangrang	..	..	200		
Total of Sutluj — Rivers	..	..	..	12,000	17,000
Total of Sutluj — Lakes, with effluence .	..	..	..	2,500	
Total of Sutluj — ,, without effluence.	..	..	..	2,500	
Karnali of Purang	..	..	..	..	1,000
All West Nari — Rivers	..	..	..	53,000	90,000
All West Nari — Lakes, with effluence . .	..	..	2500	12,000	
All West Nari — ,, without effluence .	..	..	9500		
All West Nari — Tsaka and Matsa, unknown	..	..	..	25,000	

The chief watersheds are by no means observant of the main mountain ranges, the connection or separation of basins being often determined by cross fissures and secondary ridges. This is well exemplified in *Guge;* the watershed between the *Langchen* and *Mapcha* rivers, at the S.E. corner of *Gyanima,* lies across the alluvial flat, where it is only about 15,500 feet, or still 700 feet short of its extreme elevation; and the former river, instead of making its way out along with the latter by the open gorge of Upper *Purang,* breaks through the barrier of the *Hangrang* mountains by a narrow fissure between peaks of 20,000 and 22,500 feet. Had the *Langchen* river run out by the *Serting La* (supposing the necessary fissure in the mountains at that point), and joined the *Ladak Indus* through the valley of *Hanle* (supposing the necessary changes of level), such a course would have been more normal than the existing exit to *Hangrang,* as its direction would then have been from S.E. to N.W., and the mountains about the *Serting* are more notably depressed than any other part of the whole western boundary of *Guge.* The watershed between the *Indus* and *Sutluj* has a running length of nearly 500 miles, measured from the supposed position of *Maryum La* to the N.W. extremity of *sPiti;* and some considerable portions of this are drawn across the heads of valleys transversely to the main mountain ranges.

The prevailing direction of the principal rivers is, like that of the chief mountain ranges and valleys, from S.E. to N.W.; and the main trunks of the *Indus,* the *Nubra* river, and the *Guge Sutluj,* preserve this direction through the greater part of their courses; but the cross fissures, and other irregularities in the mountain ranges, admit of frequent exceptions to this rule (as in the river of *Hembaps* and Lower *Purik,* and the river of *Sham-Zangskar*); and many of the larger affluents approximate to the same direction as the main trunks, so as to meet them very obliquely (as the rivers of *Shigar, Khapalu,* and *Yarma-Nubra,* the *Kanji Tokpo, Yuru-Wanla, Hanle,* and *Gar* rivers). Confluent branches also often meet from diametrically opposite quarters, both in the same line, and united turn off to some other direction, an arrangement which may be seen to pervade all the drainage of Upper *Zangskar* and *Lungnak.* In short, the general tendency of the main rivers is to flow from S.E. to N.W., or N.W. to S.E., and when opposed in that direction to break through the mountains till they can regain it in some other valley.

The approximate running lengths of the principal rivers are as follows:—The longest trunk of the Tibetan *Indus* is probably about 750 miles, reckoned from its lowest Tibetan point at *Acho* of Lower *Balti,* to its farthest sources in *Seng Tot* of *Nari-Khorsum.* The highest point actually attained (by myself) upon the main stream (viz., *Demchok*, on the frontier of *Ladak* and *Nari-Khorsum*) is only 550 miles or so from *Acho;* but as Moorcroft and Hearsay headed the river of *Gar*, which is well known to be a feeder of the *Indus*, 130 miles higher up, and native information so far is tolerably precise, the course of the river may be said to be positively ascertained for nearly 700 miles. Little or nothing is known of the furthest sources in *Seng-Tot* or elsewhere, and the estimate of 50 miles for the remaining course depends upon native reports and conjecture. In the lower 550 miles, I have myself mapped about the upper half of that course throughout *Ladak*, and Lieut. R. Young has done the same for the lower half through *Balti*, our surveys uniting within 25 miles or so; but the credit of the first exploration of the river belongs to Moorcroft and Trebeck, and to Mr. Vigne.

The northern trunk of the *Indus*, or river of the *Changyut* or of *Nubra* (by some called the *Shayok*), has a length of about 300 miles from its source of greatest volume in the glaciers of *Kumdan*. But the rivulet of *Gyapshan* has a further course of 50 miles to the watershed about the *Karakorum*, and makes the extreme length of this river 350 miles. If, however, there were any active effluence from the *Tso-Pangong*, and the main trunk of the river extended to the furthest affluent of that lake on the S.E., the *Kumdan* river being reduced to a northern tributary, the extreme length would be increased probably to 400 miles.

The river of *Purik* derives its importance more from the number of its branches and their volume than from the length of any of them, its longest run appearing to be hardly 100 miles from the head of the *Shingo* branch in S. *Balti*.

The river of *Zangskar* and *Lungnak* has its source of greatest volume—which is also the same for the whole *Indus*—in the head of the *Monlung* of *Lingti*, perhaps in the little lake of *Yunam*, on the N. slope of the *Para Lasa;* and the length from this source to the conflux with the *Ladak Indus* (at *Tsoksti* in Mid *Ladak*) is 180 miles, making the whole course of the Tibetan *Indus* of greatest volume 525 miles, or less than five-sevenths of the probable extreme length of the longest course; but the *Tsarap* branch of the *Lingti* river is about 20 miles longer than the others, and, measured to the furthest head of this, the *Zangskar* river will have an extreme length of 200 miles. There is also the line of dormant drainage in the lacustrine basin of N.W. *Rupshu*, connected with the *Lungnak* river by the valley of *Kyangchung* and the *Toze Tokpo;* and the furthest head of this, in the eastern affluents of the *Thogji-Chemno* lake, is just about as far (from the point of junction near *Shun*) as the source of the *Tsarap* river.

The lowest Tibetan point of the *Langchen* or *Guge Sutluj* may be placed at the conflux of the *Hangrang* river under *Namgya*, below which the joint river becomes purely *Kanori Bhotiya* or *Himalayan.* Its sources are somewhat complicated (as imperfectly known to us), lying between the *Chukar* (i. e. *White River*) from the Indian *Himalaya* on the S., the *Ser-Chu* (*Gold River*) or other streams from the mountains of *Kangri* on the N., and the effluence from the lakes *Mapham* and *Langak* on the E.; the two first being permanent, and the last partial or intermittent; and besides these, there is the dormant drainage of the *Horba* basin, which stagnates in the Lake of *Konkyu* on the extreme S.E. The *Chukar* is said to be as large as the united river of *Mensar* and *Tirthapuri*, when swelled by the melting of the *Himalayan* snow in summer; but we do not know whether it maintains this superiority on the average of the year, nor whether the intermittent contribution of the lakes be equal to the permanent affluence of the *Ser-Chu*, or other rivulets direct from the *Kangri* mountains; but the differences in both cases are perhaps small. Measured from these various sources, the length of the *Langchen* will be—from the *Darma Yankti* head of the *Chukar*, about 230 miles; from the head of the *Serchu*, behind *Kailash*, 245 miles; from the furthest affluents of *Manasarowar* on the S.E., 255 miles, which includes about 45 of the intermittent lake drainage; and from the watershed of *Maryum La*, at the S.E. extremity of *Horba*, probably about 320 miles, including 65 miles for the dormant drainage of the *Konkyu* basin.

The most important tributary of the *Guge Sutluj* is the river of *Chumurti*, which joins its right in *Rongchung*, 60 miles above *Namgya*, after running for 90 or 100 miles from the northward.

The river of S. *Rupshu Tsotso* and *Hangrang* has a course of 150 miles, from its furthest source in the head of the *Pangpok* valley to its conflux with the *Langchen* at *Namgya*: but the last 25 of this, in the valley of *Hangrang*, is only demi-Tibetan, lying a little S.W. of the line which I have assigned (though somewhat doubtfully) to the Indian watershed in this quarter; and the same character belongs to the river of *sPiti*, which joins that of *Tsotso* at the top of *Hangrang*, after a course of 80 or 90 miles from the westward. In *Rupshu*, the S.W. affluent of this river, from the N. side of the *Parang La*, though 15 or 20 miles shorter than the *Pangpok* branch, perhaps contributes as much water, because its sources are more *Himalayan*: there is also the dormant drainage of the *Tso-Moriri* basin, with an extreme course of 40 or 50 miles from the northward, and the furthest source of this is about as far as that of *Pangpok*.

The main sources of the *Mapcha* river of *Purang* lie in the N.E. slope of the *Himalaya*, N. of *Byans* of *Kumaon*; and from the furthest of these the river has a course of about 50 miles down to *Taklakhar*, which is the lowest explored point. Perhaps 20 or 30 miles more of it may be Tibetan, but this is uncertain.

In *West Nari*, neither the length of a river's course nor the area of its basin is any sure index to its volume of water. Much more depends upon the position of its sources as regards snow, and something on the nature of the soil over which it flows. From the greater quantity of snow on the Indian *Himalaya* and the Tibetan mountains next thereto, the rivers of the S. are incomparably fuller than those of the central and northern zones. Thus the Tibetan *Indus*, which has its longest course of 750 miles in the *Zhunglam*, derives its greatest volume from the *Lhogyut*, no further than 525 miles; the *Senge Sangspo*, after a course of more than 400 miles through *Nari-Khorsum* and Upper and Middle *Ladak*, probably scarce exceeds in volume the upper part of the *Lungnak* river in Lower *Lingti*, not 50 miles from its furthest sources, or the joint river of *Purik*, after running less than 100 miles; and the discharge of the whole *Zangskar* river, after running only 200 miles, and draining a basin of 6,000 square miles, mostly in the southern zone, is perhaps tenfold of the water brought by the *Ladak Indus* in double of that course, and from double the area of country, in the central districts.

Rivers of considerable volume sometimes issue ready formed from glaciers; but these are not common in *Ladak*, and exist chiefly at the heads of the northern affluents of the *Nubra Indus*, whereby the volume of that river is made little inferior to the *Indus* of *Ladak* before the accession of the *Zangskar* river. In one or two instances also they are diminished or almost absorbed by running over a sandy bed; as the river of *Hanle*, which is fuller in its thirtieth mile at *Tara Sumdo*, than in its fiftieth at *Hanle Gunpa*, and reduced almost to nothing at its debouchure in the eightieth mile at *Loma-Sumdo;* and the river of *Kumdan*, which, 15 or 20 miles below its exit from the glaciers, is almost entirely lost in sand and gravel for as much further, but then comes up again, and regains its former volume.

The fluctuations of volume arising from the melting of the snow sources are very considerable; the annual changes are felt in every river and rivulet throughout the country, but the diurnal ones become obliterated at a short distance, and hardly affect the main trunks of the principal rivers, excepting those of the *Lhogyut* with affluents joining immediately from the Indian *Himalaya* (such as the *Lungnak* of *Zangskar* and the *Langchen* of *Guge*). In the *Lhogyut* the larger streams are often fordable and unfordable alternately at different hours of the day, and these hours varying according to the distance from the snow sources. As the snow melts most during the afternoon, the rivers will be fullest about sunset, within a few miles of their sources; and a further course of 50 miles or so (according to the current) will bring the flood to the next sunrise.

The utmost irregularity prevails in the distribution of volume in the different parts of any one river; at one point a stream will be double, treble, even tenfold, of its breadth at another point a furlong distant; and the depth is equally variable, and very irregular in any one cross section; such great variations in the volume being of course compensated by equal and opposite changes in the current. In the alluvial bottoms of the principal valleys, the larger rivers generally run in broad flat beds of shingle, winding from side to side, and often subdivided by low islands, or spread into wide reticulated shallows; even the small rivulets affect this form whenever there is any flat alluvial bottom; and the condensation of the stream in a narrow rocky bed, though occurring both in the smallest and largest rivers, is comparatively uncommon.

The absolute volumes of the rivers are generally far inferior to those of the Indian *Himalaya*, and an unfordable stream is seldom to be looked for, unless in the *Lhogyut*, or in one of the main trunks of the other zones, and in summer; but numerical measurements of their actual discharge are still desiderata in both regions, and, in lieu of the former, I can only give a few loose estimates of breadth and depth, taken in crossing fords or bridges, or observed from the banks.

The *Ladak Indus*, at the highest explored part of its course in *Kakzhung*, flows through a wide flat, with a stream—often much subdivided—varying from 50 to 100 yards in breadth, and a gentle current of 2 or 3 miles an hour; and in October I found the fords here not above 2 feet deep, though in midsummer some of them exceed 3 feet. In the upper *Rong* the river is more condensed, and seldom or never fordable, being at some points not above 10 yards wide, with a most violent current, and the bottom quite hidden by the depth of water (though very clear), and the bridges vary from 40 to 80 feet in waterway. In central *Ladak* it resumes somewhat of the character it had in *Kakzhung*, being a good deal spread into wide shallows, with a moderate current: at two points, where most condensed, and unfordable, it is spanned by bridges having waterways of about 100 feet; but at *sPituk* under *Le*, it is easily fordable at all seasons, being subdivided into two streams, which in the middle of May I found each about 100 yards wide, and only 1½ feet deep, though in summer the depth is greater. Immediately below this, which is the last ford, the *Ladak Indus* becomes condensed again, and then joins the river of *Zangskar*.

The river of *Lungnak* becomes unfordable in summer, with a breadth of 30 or 40 yards, immediately below the conflux of its main head-streams in *Lingti*, only 40 or 50 miles from the furthest sources. In its lower part, after the accession of several large affluents, it is much condensed, with a deep rapid stream, and average breadth of 50 yards. On entering the wide alluvial flats in the open valley of central *Zangskar*, it spreads in subdivided streams, but still unfordable at their deepest, over a breadth of half a mile, till its junction with the river of *sTot Zangskar*. The united stream is then condensed again, and confined to a narrow rocky bed for the rest of its course through *Sham Zangskar*, its volume being at the same time much increased by several tributaries, the last and largest of which, from *Kharnak*, is itself, in summer and in the lower part of its course, an unfordable river. At *Nyerak*, in *Sham Zangskar*, Dr. Thomson found the river hemmed in between high rocky walls, only 40 feet wide, with a rapid current, and spanned by a common wooden bridge; and from this to its conflux with the *Ladak Indus*, it is impassable except on ice in winter.

In Lower *Ladak* the united *Indus* is generally condensed into a deep surging river, sometimes expanded to a breadth of 100 yards, and sometimes contracted to no more than 10, in the latter cases very rapid, and always of unfordable depth. The principal bridge here, at *Khalatse*, has a waterway of about 70 feet.

Moorcroft found some of the heads of the *Purik* river hardly fordable in the beginning of summer, and he estimated the united trunk of the river below *Khar Rul* as equal to the *Ladak Indus* under *Le*.

The northern *Indus*, which rises at once with a considerable body of water from the glaciers of *Kumdan*, is said to be almost entirely absorbed in a sandy part of the valley 30 miles below, but appears again in its full volume further down. It becomes unfordable in summer below the conflux of the large stream from *Chang-Chenmo:* thence to *Shayok* I found it from 30 to 100 yards wide, and, though flowing through a broad open valley, often condensed into a single stream, and generally unfordable in the middle of September. Between *Shayok* and *Agam* the valley is very narrow and rocky, and in the end of April I found the fords—of which there were no less than sixteen to be crossed in 30 miles—from 60 to 180 yards wide, and 1½ to 3½ feet deep; and this part of the river becomes impassable during the summer months. Through *Nubra* the river runs in a wide open flat, and is always fordable at several points, even below the conflux of the large affluent from *Yarma Nubra*, being much spread into wide shallows with a gentle current, and attaining an extreme breadth of a quarter of a mile. Below this the valley becomes contracted again, and the river more condensed and rapid, its breadth commonly from 50 to 100 yards, but under one of the bridges only 80 feet; and it is generally unfordable for the rest of its course through East *Balti*, though not without exceptions in the dry season. Some expansion occurs in *Khapalu*, and Dr. Thomson found the river fordable here in the beginning of November, with a breadth of 200 yards and average depth of 2 feet; but immediately below this it is considerably augmented by the large affluent from North *Khapalu*, and becomes condensed again into a deep unfordable stream down to its junction with the main trunk of the *Indus* at *Kiris*.

According to Mr. Vigne, the northern *Indus* is 150 to 200 yards wide close above its conflux with the central river, and this latter only 80 yards wide, but deeper; and Dr. Thomson supposes that the discharge of the two streams may be nearly equal, whilst other travellers have even considered the former to be the larger. But these are mere guesses; and, considering that the two southern branches of *Zangskar* and *Purik* alone drain a joint area as large as the whole northern basin, excluding the inactive lacustrine basin of the *Pangong*, and much moister in climate, and that the whole basin of the central and southern river is nearly three times that area, it seems pretty certain that the discharge of the central *Indus* must be very much greater than that of the northern branch.

The authorities above mentioned state the united *Indus* at *sKardo* to be from 100 to 200 yards, or often more than 500 feet, in width, and 9 or 10 feet deep in the centre, even in winter, with a moderate current; and it here receives a considerable accession from the northern affluent of *Shigar*. Mr. Vigne describes it as "a furious rapid" in the contracted gorges of *Rongyul*; and it continues probably to be deep and narrow for the rest of its course through Lower *Balti*, its volume at the same time being increased by the river of *Gilgit* and other affluents. There is no extant account of the *Indus* near its lowest Tibetan point.

The southerly position of the *Guge Sutluj* gives it a respectable volume among Tibetan rivers. The mouth of the *Chukar*, its chief southern source below the lakes of *Kangri*, is said to be forded with difficulty sometimes in summer. The main trunk of the river, a little above *Tot-Ling*, is crossed by an iron chain-bridge with a span of about 100 feet, the stream being here contracted between rocks and unfordably deep; but immediately below this it is spread out again over a wide alluvial flat, and fordable in many places between *Tot-Ling* and *Tsaprang*. I have not heard of any fords below *Tsaprang*; but there are two bridges in *Rongchung*, and a third in *Hangrang*, close above the conflux at *Namgya*, which Gerard states to have a span of 74 feet.

The western head of the *Sutluj* is rather a scanty stream in the upper part of its course through *Rupshu*, and often spread in shallows to a great breadth—at *Tronyor*, near *Chumur*, to a quarter of a mile; but it becomes condensed on entering the narrow rocky valley of *Tsotso*, and is generally unfordable in summer for the rest of its course through that district. Close above the conflux of the *sPiti* river in *Hangrang*, Gerard found it 98 feet wide, and unfordable in the middle of August, but said to be fordable with a depth of 2½ feet in the dry season. The *sPiti* river was smaller. According to the same authority, the united river of *Hangrang* is spanned by a 92-feet bridge at *Kyakhar* (v. *Shelkhar*), but in two other places measured 85 and 90 yards, which he considered its average breadth, the stream being deep and rapid.

The *Mapcha* river of *Purang* appears to be unfordable in summer under *Taklakhar* (50 miles from the furthest sources), and a little above that is spanned by a bridge of about 50 feet.

The general flatness of the valleys on the Tibetan table-land renders the fall of the principal rivers far more moderate than those of the Indian *Himalaya*, and the secondary mountain rivulets, which are more rapid, have no considerable volume; so that neither of them can compare with the thundering cataracts that form the cis-*Himalayan* sources of the Indian rivers. My measurements of the running lengths of the rivers, and of the elevations of the points taken on them, are not exact enough to give the falls with much accuracy, but those contained in the table, p. 45, may be taken as useful approximations.

It appears from the table that the fall of the main *Indus* attains its minimum (so far as known), of 3 or 4 feet per mile, in the upper part of its course, in the *Changtang*, and its maximum, of near 30, immediately below that, in the upper *Rong;* whence it subsides again, with some irregularities, to about 25 feet per mile in Central *Ladak*, 20 in Lower *Ladak*, and 15 in *Balti;* and these limits nearly include the falls of the other principal rivers, excepting in their highest or lowest parts, which are sometimes steeper than 30. The greatest fall observed in the head of any river, fairly out of the mountain ravines, was only 45 feet per mile; but there may be others steeper. In both branches of the *Sutluj*, the falls increase rapidly as they pass from the Tibetan table-land to the slope of the Indian *Himalaya*, attaining at last to maxima of 80 and 120 feet per mile, or nearly double and treble of the greatest Tibetan falls. By non-advertence to this fact, Gerard fell into great exaggerations regarding the probable elevation of the lakes of *Kangri* at the head of the *Guge* river.

Falls of Rivers in *West Nari.*

	Points of Measurement.	Elevation.	Difference of Level.	Running Distance.	Fall.	Part of River.
		Feet.	Feet.	Miles.	Ft. per Mile.	
Main Indus.	Highest explored point at Demchok of Kakshung	14,000	200	65	3	Changtang.
	Conflux of Hanle River at Loma Sumdo	13,800	100	25	4	Upper Rong.
	Mahe	13,700	1900	65	29	
	Tuna	11,800	1400	50	28	Central Ladak.
	Conflux of Zangskar River at Tsokati	10,400	1400	60	23	Lower Ladak.
	Hanudo	9,000	500 ?	25	20 ?	
	Conflux of Purik River at Morol	8,500?	1000 ?	75	13 ?	Upper Balti.
	Conflux of Nubra River at Kiris	7,500?	3000 ?	185	16 ?	Lower Balti.
	Lowest Tibetan point at Acho of Hasora	4,500?				
	Total		9500	550	17	
Northern Indus.	Exit of River from Glaciers at Kumdan	15,000	2400	95	25	Kumdan } Ladak.
	Conflux of Changchenmo River at Pholongkemo	12,600	1600	50	33	Shayok } Ladak.
	Agam	11,000	900	55	16	Nubra } Ladak.
	Khoro	10,100	1200	45	27	Chorbat } Balti.
	Siksa	8,900	1400 ?	55	25	Khapalu } Balti.
	Conflux with main Indus at Kiris	7,500?				
	Total		7,500	300	25	
Yarma Nubra.	Exit from Glacier	11,700	1400	70	20	Yarma Nubra.
	Conflux with main River in Mid Nubra	10,300				
River of Zangskar.	Valley head of Tsarap branch	15,500	1100	25	44	Tsarap } Lungnak
	Conflux of main sources in Lingti	14,000	1500	55	27	Shun-Shade } Lungnak
	Conflux of Khargya River at Purni	12,500	700	35	20	Lungnak } Lungnak
	Conflux with sTot River in Mid Zangskar	11,800	1400	70	20	Sham Zangskar.
	Conflux with Ladak Indus at Tsokati	10,400				
	Total		5100	175	29	
Purik.	Head of Hembaps branch at Matayun	10,900	2400 ?	75	32 ?	Purik.
	Conflux with Indus at Morol	8,500?				
Langchen Sutluj.	Exit from Tso Langak	15,200	1700	55	31	Kangri } Guge.
	Below Kyunglung	13,500	1100	60	18	
	Tot Ling	12,400	1200	55	22	Guge } Guge.
	Bekhar	10,800	1100	25	44	
	Shipki	9,700	1200	10	120	Rongchung } Guge.
	Conflux with River of Hangrang at Namgya	8,500				
	Total		6700	205	33	
Western Sutluj.	Head of N.W. branch in Pangpok	16,200	2100	60	35	Rupshu.
	Kyungtsang, below Chumur	14,300	4100	50	82	Tsotso.
	Conflux of sPiti River	10,200	1700	25	68	Hangrang.
	Conflux with Guge Sutluj at Namgya	8,500				
	Total		7700	135	57	
sPiti.	Conflux of Khyibar Affluent	12,300	2100	55	38	sPiti.
	Conflux with River of Tsotso	10,200				

Lakes.—The lakes (in Tibetan *mTso*) form rather an important item in the drainage system of West *Nari*, as their basins occupy in all probability not less than 12,000 square miles (and possibly much more if the unexplored parts of the north-east be lacustrine), the whole of which is exclusively in the *Changtang*, where the flatness of the general level and absence of deep fissures favour such accumulations of the drainage.

The largest of all the lakes is the *Pangong*, which receives half the drainage of *Tanktse* of *Ladak*, and all of *Ruduk* of *Nari*, and, were it not a lake, would be the longest trunk of the northern branch of the *Indus*. Its basin has a mean length of about 110 miles directed S.E. and N.W., with a breadth of 60, and area, therefore, of 6,500 square miles. Though not actually explored to the whole of this extent, it is not likely to be less, and may be very much more, including dormant drainage to the eastward.

The lake itself has a running length of 100 miles probably (about half of which is actually explored), and an average breadth of only 3 miles or so, making the area 300 square miles. The middle part of the length is directed nearly E. and W., the N.W. quarter being turned northward, and the S.E. probably southward.

The character of the containing banks is for the most part very steep and rocky, scarcely leaving a footpath along the water's edge, but exceptions to this occur in several places; all along the S.W. side of the northern horn there is a flat or shelving alluvial bank, sometimes a mile wide, giving site to the agricultural hamlets of *Pangong*, where mountain rivulets admit of them; and there are also said to be wide plains open to the lake in the southern horn, at *Ruduk* in the S., and at *No* on the N. side, but these formed in whole or part by the mouths of the lateral valleys opening into the immediate basin of the lake.

The plain of *Ot*—which is 7 or 8 miles long, with an extreme breadth of 3 miles, on the N. side, about 10 miles W. of the lake's mid length, at the mouth of the *Chang-Parma* valley—encroaches altogether on the proper bed of the lake itself, which is here singularly contracted, for 2 or 3 miles, into a narrow channel like a large river, having a minimum breadth of 50 yards of unfordable depth.

The lake has at least eight permanent affluents; three only are actually explored in the N.W. half, but we have precise information of five more in the S.E. Of the former, the longest has a run of 40 or 50 miles (through *Chang-Parma*), but the shortest, of only half that course (from *lData*), has the largest volume; of the latter, it seems probable that one passing *Ruduk* (from *Chagang*, &c.) has a run of 60 or 70 miles, and another, debouching at *No* (from the N.), is said to be hardly fordable at its mouth in summer.

The *Pangong* has no effluence whatever; but there is an open valley connecting the N.W. end of its basin with that of the

Tanktse affluent of the *Nubra* river, *viâ Muglib*, the length of dormant drainage between the edge of the lake near *Jaktil*, and the first appearance of the *Muglib* rivulet in a scanty spring at *Wangtong*, being 7 or 8 miles; and the spring, being sub-saline, is considered by the Tibetan inhabitants as a filtration of the lake water. The watershed across the head of the valley is almost imperceptible, but lies probably at *Donzho Lhato*, only a mile from the lake, and scarce 100 feet above its present surface. The present level of the water is about 13,400 feet.

All along the banks of the lake there is a well-defined zone of horizontal watermarks, extending to a height of perhaps 70 feet above the present surface, formed both by calcareous concretions, and by erosions on the foot of the marginal rocks, corresponding marks being also visible in parts of the alluvial shore; and the uppermost of these lines no doubt marks the level of the existing watershed at *Donzho*. The gradual subsidence of the lake is established by further evidence. The plain of *Ot*, which is raised only 10 or 12 feet above the present water-line, consists of fine earthy strata full of small shells, which are very perfect and unaltered, if fossils at all of the most recent sort, and some of them closely resembling (if not identical with) a small species still living in the *Tso-Rul*, though none were observed in the water of the *Pangong* itself. Although the inhabitants of *Pangong* have no tradition regarding the origination of *Ot* by the subsidence of the lake, its gradual desiccation in the present (*i. e.*, human *v.* historic) era seems probable; and I was informed by a Tibetan, who had visited the lake in company with Moorcroft in 1821, and again with myself in 1848, that the water had receded perceptibly from the encamping ground at *Jaktil* during these 27 years, which, as the shore there is flat, might have been done by a very slight subsidence of the water.

The water of the *Tso-Pangong*, in the greater part of the lake, is very salt (not the salt of the sea, but Epsom salts, or some other compound of Soda), and the solution seems doubly concentrated in the contracted channel at *Ot;* but at the S.E. end of the lake it is said to be quite fresh and potable, which no doubt arises from the greater affluence of mountain streams in that quarter.

The *Tso Rul* and *Tso Shaldat* are two small lakes, situated in a flat valley, which connects the *Chushul* valley with that of *Ruduk*, parallel to the *Pangong*, and a few miles S. of its middle part, with a length of 50 or 60 miles, and average breadth of 1, uniformly elevated about 14,400 or 14,500 feet, and with no sensible watershed, its western and longer portion forming the basin of the *Tso Rul*, and a few miles at the E. end that of the *Shaldat*. The *Tso Rul* is about 15 miles long E. and W., and 1 mile broad, with a single scanty affluent 10 or 12 miles long entering its E.

end, and one or two small springs of fresh water rising in the hill foot at its margin; the lake water being salt, like that of the *Pangong*, though in a less degree, and without any effluence, though the basin is quite flat and open towards the *Chushul* valley. The *Shaldat* is only 3 miles long, and at the utmost 1 wide, with a very scanty and intermittent affluent to its S.E. corner, no effluence, but equally open to the *Tso Rul* or to *Ruduk*, and its water not so salt as the other, but still undrinkable. Both of these lakes are surrounded by signs of modern subsidence like the *Pangong*.

The general system of the *Tso Pangong* may be finally explained by saying, that a very slight elevation in the country about *Kesar Gidpo* (between the two small lakes) would suffice to send out the *Shaldat* by *Ruduk* and the *Tso Rul* by *Chushul*, into the great lake, and an elevation of one or two hundred feet at *Nertsong* (at the S.E. end of the *Pangong*), with an equal depression at *Donzho* (at the N.W. end), to drain off the whole into the *Nubra Indus* by *Tanktse* and *Shayok*; but if the present state of things last long enough, the two small lakes will become extinct, and the *Pangong* divided into two lakes, separated by an isthmus at *Ot*.

The lakes of S.E. *Nari*, at the head of the *Langchen Sutluj*, drain in all a basin about 90 miles long S.E. and N.W., and perhaps 50 wide, making an area of 4,500 square miles. The S.E. half of this, comprising the valley of *Horba*, contains the *Tso Konkyu*, lying immediately under the W. foot of *Maryun La*; it has not been explored by any English traveller, but my Tibetan informants represent the lake to be long and narrow like the *Tso Rul*, but larger—perhaps 20 miles long and 2 wide, or 400 square miles in area—with no effluence, yet no high ridge between it and the lakes of *Kangri*, and certainly belonging to the same system of drainage; the water salt; and the elevation probably between 15,500 and 16,000 feet.

The north-western half of the same basin comprises the greater part of *Kangri*, and contains the *Tso Mapham*, or Indian *Manasarowar*, and the *Tso Langak* (or *Oma Tso*, i. e. *Milk Lake*), the *Rakas Tal* or *Rawan Rhad* of the Indians. Detailed accounts of these have been already published in the volumes of the Calcutta Asiatic Society, and I here confine myself to their more prominent geographical features. They lie together in a large southern recess of the *Kangri* valley, S. of the mountains that impart this name to the whole district and include the notable peak of *Tisé*—the Indian *Kailash*—and N. of the still more gigantic *Gurla*; they are parted off by detached groups of inferior mountain from *Horba* and the *Tso Konkyu* on the E., and from the head of *Purang* and *Gyanima* of *Guge* on the W.; and a spur of low hills—partly alluvial, I believe—projecting from the foot of *Gurla*, divides the two lakes by an isthmus 4 or 5 miles wide, also partly shutting off the eastern one from the wide flat valley on the N. They are both upon the same level (within a few feet), at an elevation of 15,200 feet; and both of the purest fresh water, in spite of much saline exudation in many parts of their banks.

The eastern lake *Mapham* is of oblong or oval shape, 15 or 16 miles long E. and W., and 11 or 12 broad N. and S., making an area of 150 square miles. It is known to have several affluents from the high snowy mountains that flank it on the N. and the S., though the English explorers, confined as yet to the W. side, have not actually crossed any of them; and one of them, from the *Kangri* mountains on the N.E., is said to form a small fresh-water lake, called the *Tso Khurgyal*, shortly before its debouchure into *Mapham* (the high road to *U-Tsang* going between the two lakes). Its effluent runs through an opening in the hilly isthmus into the western lake: I found it a swift stream 100 feet broad and 3 deep in October; and its exit from the lake was seen by Mr. J. E. Winterbottom and my brother, Richard Strachey, from a height a few furlongs off; though Moorcroft crossed this very place without finding the stream, even in August, the time of highest flood—whether from its being then dry, or from its percolating through a bar of shingle close to the lake, as asserted by some of my native informants, is not apparent.

The western lake *Langak* has probably about the same area as *Mapham*, but an irregular shape, with a length of 20 miles N. and S., and extreme breadth of 15, measured obliquely across its S. end, the N. end narrowing off to a point; at the S. end there are two or three small rocky islands, the only known instances of such in the lakes of West *Nari*. My brother and I have between us completed the circuit of this lake, so as to leave no doubt of its affluents and effluents; the former comprise the effluent of *Mapham* above noticed, two streams from the *Kangri* mountains on the N.E. as large as that, and a much smaller one from the valley of *Jungba* in the hills on the S.E. The effluence to the *Langchen* river is from the N. point, which I crossed however in October without finding any running stream, or any marked channel for one, though the flatness of the ground, its partial inundation in shallow pools, and obvious descent of the level towards the river, entirely corroborated the native accounts of an intermittent effluence in seasons of flood. It is worthy of notice that one branch of the *Jungba* valley is connected with valleys draining into *Gyanima* of *Guge*; and the level of both so nearly flat, and the dividing watershed so insensible, that the depression of a few feet at this end, with an equal elevation at the N. point of the lake, would certainly turn the effluence into the *Chu Kar*, through the plain of *Gyanima*, where there now exists a small lake 5 or 6 miles long and 1 in extreme breadth.

The basin of the *Tso Moriri*, in South *Rupshu*, connected with the *Sutluj* through *Tsotso* and *Hangrang*, consists of two separate parts, the one of permanent and the other of intermittent drainage. The former, to the N.E., has a length of 28 miles N. and S., with a breadth of 25 miles at the N. end, but only 3 or 4 at the S., making an area of about 400 square miles; and this contains the lake at its S. end: the latter joins at the S. point of the former, from the S.W., having nearly an equal area, with a length of 25 miles N.W. and S.E., and mean breadth of 17; and including half of this the whole basin may be reckoned equivalent to 600 square miles of permanent drainage. The surrounding mountains are not above 19,000 feet, and still less in the southern part, and generally sloped off to the lake with a margin of broad shelving bank. The lake itself has an oblong shape, with a length of 14 or 15 miles N. and S., breadth of 3 or 4, and area of about 45 square miles; its elevation is 15,200 feet, and its water subsaline. The longest permanent affluent is that of *Gyang* and *Tsakshang* from the N.W., with a length of 25 or 30 miles; that of *Karzok* on the W. is much smaller, and two or three others are insignificant summer rivulets.

The intermittent affluent which joins the S. end of the lake, after a course of 40 or 50 miles in two branches from the westward, is the same with the *Pangpok* branch of the *Rupshu* river. This stream sends half its waters into the lake and half to join the *Parang* branch of the river at *Nurbu-Sumdo*, sometimes both at once and sometimes alternately in different years; an instance of distomosis which, if insignificant for the smallness of its scale, is remarkable as the only one that I have ever seen or heard of in all *West Nari*, or the Indian *Himalaya*. The mouth of the *Pirse* valley (which is the eastern continuation of *Pangpok*) is contracted into a narrow rocky gorge, opening at right angles into the W. side of the *Leptra* valley (which is the southern prolongation of the lake's immediate basin), where the latter is 3 or 4 miles wide; and though the bottoms of both are generally very flat, a large fan-shaped alluvial delta spreads out of *Pirse* right across the breadth of *Leptra;* and the middle part of this bank, being as usual the highest, forms a low flat ridge, so exactly medial to the debouchure of the river that the stream wanders over it for some way, in widely subdivided shallows, before ultimately taking to one or other slope of the delta, or to both; and once parted the two streams cannot unite again, because the medial rise of the bank forms a complete though low bar across the whole bottom of the *Leptra* valley. A change of a very few feet in the relative levels in some parts of the divaricating alluvial bank, would suffice to put an end to the distomosis, and determine the flow of the *Pirse* river either wholly into the lake, or wholly into the river of *Rupshu*.

The *Tso Moriri* has no effluence; but an elevation of 100 feet at its N. end, with an equal depression at the S., would send it all into the *Rupshu* river, along with the river from *Pirse*, through *Leptra* to *Nurbu-Sumdo*. The old idea that the lake actually formed the source of the river, obtained by Herbert and others from distant inquiries, and still figuring on most English maps, though not strictly correct, is thus founded on geographical facts, and a careless observer might mistake the southern branch of the *Pirse* river for an effluent of the lake.

I saw no signs of desiccation in the lake, nor is the existing watershed across the *Leptra* valley consistent with any considerable height of water above the present surface.

The lacustrine basin of North *Rupshu*, connected with the system of the *Lungnak Indus*, is also divided into two parts. The eastern one, containing the lakes of *Tsowar*, is a triangular area of about 200 square miles, the lakes being situated near the middle of its W. side, where there is an opening through low hills communicating with the western division of the basin at *Rokchung*. This latter has an equal area of 200 square miles, contains the valleys of *Tanglang Rok* and *Kyangchung*, and opens at its south end into the *Toze Tokpo* branch of the *Lungnak* at *Pangtik*; it has no permanent rivulets of any size, and but one or two insignificant ponds often quite exhausted in the dry season.

The main lake of *Tsowar* has an area of a few square miles only; it has one or two small mountain affluents, the chief of which from the southward expands a little before its debouchure into a pond or small lake of fresh water. It has no effluence, though the opening for it exists as above mentioned: and Dr. Thomson observed upon the surrounding hill-foot a line of ancient water-marks, corresponding with the low valley watershed, and indicating a desiccation of the lake, about 150 feet above the present level. The elevation is somewhere about 15,000 feet; the water of the larger lake-salt, and a small part of the marginal concretions sometimes found to be edible sea-salt.

The name *Tsowar* signifies *Between the Lakes*, and is applied to the summer camp of the *Rupshu* shepherds in that situation. *Thogji Chemno*, the *Great Benevolence*, is the name of the *Lha*, or local deity, who has his *Lhato* or dedicated turret, on the N. side of the larger lake, which itself must be called by the same name for want of any other.

In the preceding notice of the lakes I have made no allusion to their depth, because it is not known in any instance; but with reference to the general formation of the valleys in which they are embedded, it may be conjectured to range within two or three hundred feet.

The following peculiarities are worthy of remark, as seeming to pervade the whole system. Every one of the Lakes has a connexion by open alluvial valley with the general river drainage of the country: none of them can ever have been much (or 100 vertical feet) fuller than they now are: they are not to be found below 14,000 feet: some of them are drying up without any effluence, under the sole effect of evaporation or subterranean filtration: and those with an effluence are invariably fresh-water, those without it invariably salt.

In addition to the lakes above described, there are several small ones, all fresh-water, formed by the stagnation of rivulets in flat valleys or ravines, generally at high elevations; but they have no geographical importance, and are often mere duck-ponds, though interesting as objects of topography or landscape, and seldom without Tibetan names. The largest that I have heard of is the lake of *Kyung*, said to be 3 or 4 miles round, in the ravine of *Nidar*, entering the left of the *Indus* in Upper *Ladak*, probably at a height of 15,000 feet; and the most curious that I have seen, is a little pond a few furlongs round, completely insulated in a crater of low serpentine hills that protrude through the alluvial bottom, in the valley of *Yarma-Nubra*, between *Tiritsha* and *Panimik*; but this is in the Rong. at a height of only 10,700 feet.

Glaciers.—Glaciers are well known to the Tibetans under the name of *Kangri*, i. e. *Iceberg*, which is also loosely used to denote any high mountain covered with perpetual snow or *nevée*; but, as might be expected in a country of such little snow, they form rather the exception than the rule, and are not commonly to be found away from the Indian watershed.

In the central part of *Ladak* the only instances I met with were in the very lofty range dividing *Pangong* from *Tanktse* Proper, where several small glaciers of the incipient class fill the deeper of the top ravines below the general coating of nevée, but yet so elevated, that from my camp in the valley of *Ko*, at 17,800 feet, the lowest of them still seemed 1,000 vertical feet above me. From the shore of the *Pangong*, about *Man* and *Merak*, the ends of some of these may be seen hanging upon ledges of the mountain overhead, and their true character is sufficiently attested by the Tibetan inhabitants, who (incurious as they generally are in such matters) have observed their progressive motion and *éboulement*, as evidenced by fragments of ice falling to the foot of the mountain. Symptoms of terminal moraine may also be seen descending through the lateral ravines in the valley of *Muglib*;

and the head of the *Ko* valley above mentioned appears to be composed of ancient moraine rather than alluvium.

But the chief reservoir of Tibetan glaciers seems to be in the S. face of the Turkish watershed, which the joint observations of English travellers and native reports prove to be full of them, and many of the first class both for size and formation. The main trunk of the *Nubra* river issues from two of these, at a place called *Kumdan* (by the Turks *Chongtash*, i. e. *Willow Boulders*), where the position of the glaciers is such as to cause most devastating *débacles*, which I shall describe further on. Dr. Thomson visited this place, and has given an account of the glaciers.

I myself found the river of *Yarma-Nubra* issuing full-formed (being 50 yards wide, with an extreme depth of 1½ feet, and very rapid, in the beginning of October) from a large glacier, entirely occupying the head of the valley and (so far as Tibetan information goes) rendering it impassable. I estimated the breadth of the glacier, in its lower part, to be ¾ of a mile; and its length is such that, after ascending perhaps 2 miles, I was unable to see the head of either of the two branches into which it is divided 4 or 5 miles above the lower end. The thickness of ice seemed at least 200 feet. The elevation of the lower end is about 11,700 feet, which is probably 8,000 feet below the mean level of perpetual snow in these regions (viz., 20,000 feet). Wild juniper trees grow all about the hill sides along its lower part, and though the valley is not actually inhabited for several miles below, a fragment of alluvial bank on which I encamped might easily have been converted into a corn-field in immediate proximity to the foot of the glacier. The glacier is remarkable for the extreme flatness of its level, and the abrupt contact of its ice with the steep walls of the granite mountain on both flanks, in which points it follows the character of the retaining valley bottom. The surface also is very free from moraine, often exhibiting the pure clean ice; and although passable with care and toil, in its lower part at least, the crevassation is so excessive that the whole glacier seems to consist of an aggregation of great lumps—each as big as a house, and jammed together in the utmost disorder—rather than a continuous mass with cracks in it: a formation which, I think, must be ascribed less to the disruptive inequality of motion than to the action of surface melting, which produces a network of rivulets that cut grooves in the ice, soon deepening into chasms, which often insulate and undermine great blocks, till they subside or tumble over into new forms of confusion. Glaciers, however, to be well understood, require much experience and specific study, as well as days and weeks of exploration; and my own hasty observations on the *Kangri* of *Yarma-Nubra* may be superseded by more elaborate inquiries hereafter.

The *Tulumbuti* affluent of the *Yarma-Nubra* river also rises from glaciers, which are passed on the summer road to *Yarkend,* upon the S.W. of the *Saser La* (as mentioned by Dr. Thomson). Mr. Vigne found several glaciers in the *Shigar* and *Khapalu* valleys, aligning with those of *Kumdan* and *Yarma-Nubra;* and the native travellers between *Yarkend* and *Balti* testify to a very large one upon the Turkish watershed, at the head of the *Braldo* branch of *Shigar,* which forms a serious obstacle to this route, and gives the Pass its Turkish name of *Mustag,* i. e. *Iceberg.* Mr. J. E. Winterbottom and Lieut. R. Young found another still farther to the N.W., beyond the Tibetan frontier, in the northern head of *Gilgit.*

In connexion with glaciers, I must notice the occurrence of certain permanent beds of frozen snow, or Valley Nevée, which I met with in two or three places of the *Changtang* in the middle of summer, when the snow-line was thousands of feet higher, and in situations where they could not be ascribed to avalanche or to unusual shade (as in the low summer snow-beds of the Indian *Himalaya*). The Tibetans call these snow-fields *Dar,* and assert them to be permanent, and constant from year to year in the same places. The largest that I met with were in a wide sunny valley, among low snowless mountains; and all of them in the beds of rivulets where most flat and marshy; the snow always hard frozen, with a very irregular surface, such as often occurs in glaciers or nevée; and the beds more or less broken up into patches, with every variety of thickness up to 3 or 4 feet. I have not been able to form any distinct opinion of these *Dar,* whether they have in fact any peculiar cause and constitution of their own, or whether they are mere sporadic extremes of the lowest perpetual snow; but to help the resolution of these questions by future observers, I here record the instances which attracted my own notice.

11th August, 1848.—In *Chang Parma,* of *Pangong,* a *Dar* about 3 furlongs long and 1 wide, and 2 or 3 feet thick, at *Mitpal Yogma,* the valley bottom elevated 16,100 feet and tolerably open. *17th Idem.* A small *Dar* at *Phea Tot,* bottom elevated about 16,500 feet, and rather contracted. From two or three mountain passages in the vicinity, the mean snow-line was estimated about this time to be nearly 20,000 feet. *21st June,* 1849.—In South *Rupshu,* passed two large fields of *Dar,* covering several acres, in places 4 or 5 feet thick; one at *Manechan-Sumdo,* 15,800 feet, and the other 3 or 4 miles above it towards *Pangpok,* about 16,000 feet; the valley bottom being a mile wide, and exposed to the sun all day, and the surrounding mountains low and snowless; the mean snow-line estimated, from a neighbouring mountain passage the day before, at 17,500 feet, but a fortnight later up to 19,000 feet.

Débacles.—The great valley of the *Maryul Changyut* appears to have been subject, from time immemorial, to devastating floods rising from débacles in its north-eastern head. The earliest traditions of them take the form of mythology, and magnify them to a scale fit for the giants and Titans, by confounding their effects with the extant ruins of the ancient alluvial era. Thus, the Tibetan inhabitants of these valleys relate that the oldest and greatest of the floods was effected by three of the "*Lha*"—i. e. *Local Deities*—to whom they still erect towers, and offer sacrifices: *Rangisha* of *Sakti* let loose the waters in the head of the valley at *Kumdan; Zangnam* of *Yarma-Nubra* dammed them up at the gorge of *Khoro* between *Nubra* and *Chorbat;* and *Koyak* of *Nubra* drained them off again by breaking *Zangnam's* dam; and the supposed marks of all these operations are to be seen thousands of feet above the present level of the river.

The oldest flood of which I could get any credible matter-of-fact account occurred only about seventy years back, and was witnessed by the oldest inhabitant of *Shayok,* then a young girl, now an old woman of eighty, who described it to have been on a great scale like that of 1835: between these, there seems to have been no flood of any note.

The one last mentioned occurred on the seventeenth day of the fifth moon of the "*Sheep Year,*" which was some time in June of 1835. As the valley is not permanently inhabited above *Shayok,* and none of the *Yarkendi* caravans happened to be at *Kumdan* at the crisis of the débacle—though there but a day or two both before and after it—there were no eye-witnesses to its actual commencement; but the *Yarkendi* travellers have been observant enough to ascertain and describe the cause, and their accounts have been confirmed by Dr. Thomson's visit to the spot in 1848.

The débacle originates in two glaciers a mile or two asunder, the upper one of huge dimensions—perhaps a mile wide—coming from lateral ravines on the proper right of the main valley, so transversely that in process of time they slide right across the bottom of the latter, and after abutting upon the steep wall of rock upon its E. side, are thrust upwards by the pressure from behind till the ice reaches a vast height, in the upper glacier—perhaps 700 vertical feet, above the valley bottom. This obstructs all the drainage of the upper valley—probably a running length of 50 miles, and a basin of several hundred square miles—and perhaps some portion of the glacier-river itself, till the waters become accumulated into a large lake between *Kumdan* and *Gyapshan;* and under the weight and solvent power of this body of water, together with the continued pressure of the descending glacier, the lower end of the ice is at length burst asunder with a sudden and violent disruption, discharging the accumulated water all at once in a torrent of immense volume and rapidity, hurrying along with it vast quantities of the overturned ice and moraine.

Travellers who passed *Sultan-Chuskul* a few days after the débacle, found the valley full of blocks of ice as big as houses, and so spread with heaps of soft mud that it was impossible to get up to the *Saser* Pass into *Nubra*, though the flood-water had all run off. Only ten or twelve years before the débacle, the foot of the glacier was not within a quarter of a mile of the eastern mountain, and the *Yarkend* road lay through the open passage; but within that period the jamming of the glacier, and damming of the *Gyapshan* water, had rendered the valley impassable above *Kumdan*, and obliged the road to take another line to the E. of these obstructions; nor was the disruption of the ice in 1835 sufficient to restore an opening practicable for the road, or offering security against an early recurrence of the débacle, the upper glacier being still impassable when Dr. Thomson visited it in 1848.

Shayok, the highest permanent habitation of this valley, is distant about 100 miles from *Kumdan*, and the village being situated on a bank 200 or 300 feet above the river, is much beyond the reach of the greatest known floods. That of 1835 is said to have passed it by night, and to have mostly run off by the next morning; though, if the statements which the villagers made to me were true, the water must have risen nearly 50 feet where the bottom is about a mile wide. It passed Mid-*Nubra*, which is 60 miles below *Shayok*, some time before daybreak, and had run off by noon of the same day. It here swept away all the lower half of the village of *Lagzhung*, with ten out of the twenty-two houses, and all their inmates, men and cattle; and the village of *Deskit* suffered equal loss in its low-lying quarters. These were the only two villages in all *Nubra* low enough to be touched by the flood. At *Lagzhung* I found fragments of the old fields standing 5 or 6 feet above the low sandy waste of the river flat; and as the remaining part of the village is only about as much higher, the flood could not have risen here above 8 or 10 feet; but attaining even that height, it must have spread into the mouth of the *Yarma* valley several miles.

In *Chorbat* of *Balti*, from 30 to 80 miles below Mid *Nubra*, and chiefly in the further part of that distance, no less than 150 farms were swept away from the low-lying lands in the river-bed, but with less destruction to human life, as the flood passed here between day-break and sun-rise, and the inhabitants, warned by the roar of the approaching torrent and trembling of the earth, had time to effect their escape, with the loss of lands, houses and cattle. The main part of the flood-water is said to have run off from *Chorbat* in a few hours, and nearly the whole of it in two days The more destructive ravages did not extend below *Chorbat* or Upper *Khapalu*, but the inundation was dangerously great down to the debouchure of the northern valley at *Kiris*, in Mid *Balti*. Mr. Vigne has made some mention of its effects upon the united Indus at *sKardo*.

Throughout the 300 miles, from *Kumdan* to *Kiris*, the flood committed great havoc with the natural brushwood of tamarisk, hippophae, and willows, that grew abundantly in the river-bed; many square miles of valuable *Tsok* (*i. e.* thorn shrubbery) were thus destroyed in Mid *Nubra*: and all the way from *Agam* to *Kumdan* the *Yarkendi* traveller has still to lament the conversion of the pleasant little thickets and grass-plots that once gave him shelter and forage into barren wastes of sand and gravel. Further mischief was done by the violence of the torrent ploughing up deep holes in the softer parts of the river-bed, some of which still remain to render the stream unfordable, and increase the difficulties of the eastern route to *Yarkend*. The most dangerous mountain-path I ever crossed in my life was necessitated by one of these unfordable depths 10 miles above *Shayok*. For 150 miles below *Kumdan* the valley bottom was strewn with fragments of the glacier, the largest of which were several years in melting away: in 1848 I myself saw the residual moraine of some of them, half-way between *Agam* and *Shayok*.

The Tibetan peasantry are not exact enough in their mode of reckoning time to admit of any precise determination of the velocity of this flood; but I think it may be assumed that it did not pass *Shayok* before midnight, nor *Khapalu* after 6 A.M., making not less than 150 miles in 6 hours, or a rate of 25 miles per hour; but it may have been much more rapid.

In spite of so great a cataclysm as this in 1835, a débacle from the *Kumdan* glaciers occurred again in the next "*Hog-Year*," or 1839; but this was of much less extent, and passed *Nubra* at mid-day; and as all of the villages liable to inundation had been destroyed by the former flood, merely carried away the cattle and herdsmen that happened to be out at sea in the river flat; and in the upper part of the valley it did a little good perhaps, in levelling some of the mischievous pits and mounds that had been ploughed up by the great flood of 1835.

Mr. Vigne has associated these Tibetan débacles with some extraordinary floods of the *Indus* observed in the plains of India; but the two are, I believe, quite incommensurate both in scale and dates of occurrence, and the latter well known to have been caused by landslips altogether below *Balti*.

Subterranean.—No traces of volcanoes have been seen or heard of in any part of West *Nari*.

Hot springs occur in all quarters. They have been noticed by English travellers in more than a dozen different places, which is probably not near all that exist; these are generally very scanty, never up to the boiling point at their issue, and sometimes accompanied with gaseous exhalations.

Earthquakes are not unknown: I myself experienced a slight shock of one in Central Ladak (at *Chimra*, evening, 1st of June, 1848), and Dr. Thomson mentions another at the head of the Northern Indus (at *Morgo Chumik*, near *Kumdan*, 23rd of August, of the same year), but the state of several old buildings in *Ladak* proves that none of any intensity have occurred there for the last 250 years at least.

Climate.—Temperature.—The mountainous contour of the Tibetan Table-land gives rise to a great variety of temperature-climates, corresponding with the various elevations of the country, besides what differences arise from other causes, such as northing or southing in latitude, expanse or contraction of valleys, exposures to sun and wind, &c. The diversities of climate thus produced are in fact innumerable, and I can only attempt to notice one or two of them, chiefly in *Ladak*, to which region my own observations were mostly confined.

The town of *Le*, elevated from 11,800 to 12,000 feet, may be taken in point of elevation as an average of the inhabited valleys of all *Ladak*; but its site on a south face of granite rock, backed by high mountains on the north, and exposed to a greater expanse of south sky than almost any other place in the *Rong* country, raises its temperature probably above the average of equal elevations.

Continual frost, lasting in the shade throughout the day, is nearly coincident with the winter quarter in the town of *Le*, but in less favoured localities it begins much sooner even at much lower elevations. Thus in a north ravine of *Chorbat*, elevated 12,000 feet, the temperature was only 28° at 2 P.M. on the 4th of November, 1848. Constant frost set in on the 13th of the same month at *sKurbuchan* of Lower *Ladak*, elevated only 9,700 feet, with a south aspect; and from this up to *Le*, during the next ten days, I found the side rivulets (all with a south exposure) little better than cataracts of ice, and the *Indus* rapidly filling with sludge. In my lodgings at *Le* the frost was hard and constant from the beginning of December to the end of February: in any part of the house away from a fire, water froze speedily, and, if left so, remained in the form of ice throughout the season; and during the coldest parts of the winter, water, ink, milk, &c., would freeze in spite of fires, and close to them; the house, however, was a large one, kept in the Tibetan fashion, well exposed to the open air, and not many degrees warmer. In this situation

I never saw the thermometer below +2° Faht.; but as this was at 9½ A.M. (10th of February, 1848), and the exposure of the instrument by no means perfect, it cannot be taken as the true external minimum: the weather, however, about that time was much colder than the average of the winter. On the whole, the winter temperature of *Le* town may be said to range between 0° and 30°, to be colder than usual when below 10°, and warmer than usual when above 20°.

The rivulet of the *Le* valley is one mass of ice throughout the winter months. The *Indus*, below the town, at a height of 11,000 feet, freezes over during the same period; but owing to the great rapidity of the current, and to springs in the river-bed with a subterranean temperature, the ice is by no means continuous or durable. All the larger rivers may be reckoned to freeze over in winter down to the lowest level of *Ladak*, or 8,000 feet, though, for the causes above mentioned, their ice is often very much broken. Dr. Thomson found the *Indus* frozen over at *Khartaksho* of *Balti*, about 7,700 feet, on the 19th of December, 1847.

The rise of temperature about the vernal equinox is great and sudden in the town of *Le*, owing to the sun then striking on a multitude of brick and stone walls and granite rocks that his rays cannot reach in winter. In the fields, the moderate night frosts of spring appear to cease in the middle of April, although thin films of ice may be seen a month later about the shallows of the *Indus* 1,000 feet below—an effect, no doubt, of greater nocturnal radiation in a more open situation.

The summer temperature of *Le* is very hot if the direct effect of the sun's rays be taken into account, but quite the reverse as regards air temperature in the shade. The highest temperature I ever observed there was (at 3 P.M. 5th of July, 1848) 69° in good shade and wind, in the fields below the town, at a height of 11,800 feet, the thermometer just before sunrise on the same day having been down to 53°; and on the 23rd of the same month, at a station about the same elevation a little higher up the *Indus*, the temperature rose to 71° at noon in perfect shade and wind: as the weather had been unusually hot about the time of these observations, I consider that the ordinary maximum of perfect shade temperature at a height of 12,000 feet, in Central *Ladak*, does not exceed 70°. But perfect shade is very rarely to be found in this land of bare rocks, and the best attainable in most places will admit so much reflected heat as to raise the thermometer 8 or 10 degrees. Under such circumstances I once saw the thermometer up to 81½°, at 1½ P.M. on the 1st of September, 1847, at *Pok* of Lower *sPiti*, elevated 11,600 feet, the latitude, however, being 2° S. of *Le*.

Night frosts begin at *Le* in the middle of September probably.

In the valley of Mid-*Nubra*, at an elevation of 10,500 feet, in the autumn of 1848, the first ice was to be seen in the end of September. By the middle of October, in the same locality, the night frosts became very sharp, and, after that, continued increasing whilst I was descending to lower elevations, and by the end of the month, in Lower *Nubra* and *Chorbat*, the day maximum, in cloudy weather, was little above the freezing point, at heights of 12,000 feet.

My acquaintance with the lower parts of *Ladak* is so small that I cannot give any separate account of climates warmer than *Le;* but long excursions in the *Changtang* enable me to add a few particulars regarding the colder climates of greater elevations.

Elevations not above 14,000 feet appear to be exempt from night frost for the greater part of the summer quarter. At heights of 15,500 feet it freezes probably every night of the year. The greatest height at which I ever passed the night without finding ice in the rivulets in the morning was 15,300 feet, viz., at *lData* of *Pangong* on the 2nd and 3rd of August, 1848; but during the next two weeks frost occurred every night, at other places of equal or greater elevation in the same district. In 1849, when encamped by the *Tso Moriri*, at a height of 15,200 feet, I experienced frost every night from the 23rd of June to the 3rd of July; and in travelling through the uplands of *Hanle Chumurti* and *Guge*, during the next three weeks, at elevations generally exceeding 15,000 feet, I found the rivulets frozen nearly every morning, excepting in the ravines of Central *Guge* below that height.

The maximum shade temperature, at heights of 15,000 feet, is probably about 60°; but as good shade is rarely forthcoming in such places during the middle of the day, the thermometer will often be higher. I once saw it up to 70°, at 1¼ P.M. on the 11th of September, 1847, at *Tronyor* of *Rupshu*, elevated 14,900 feet. The day temperature continues mild enough at these elevations till the autumnal equinox, after which it decreases very rapidly, and the frost becomes constant, sooner or later in the month of October. At *Hanle Gunpa*, elevated 14,500 feet, the thermometer still rose above 40° during the afternoon in the first week of October, 1847, but there was a good deal of reflected heat here from the walls of the building. During the second week of the same month, in the valley of *Kakzhung*, averaging 14,000 feet, the temperatures were between 35° and 40° during the warmest part of the day, nor was there any permanent ice in the *Indus;* though the night temperatures were very low, the thermometer ranging from 6° to 11° at sunrise--an instance, I suppose, of increased nocturnal radiation in an open plain. The greatest cold which I ever myself experienced in *Ladak* was in crossing from *Chorbat* of *Balti* to *Hanu* of Lower *Ladak*, in the beginning of November, 1848, the temperature at the N. foot of the Pass, elevated 14,600 feet, not exceeding 18° during the day of the 5th, and falling to 1° at sunrise the next morning, being about the same as one of the coldest winter days at *Le*.

I had no personal experience of the winter at these heights, and the only information I can give on the subject is, that the lakes, elevated from 14,300 to 15,200 feet, are completely and strongly frozen over for three or four months in the year. As the freezing of large bodies of water is rather a sudden operation, following—not accompanying—the gradual depression of temperature, whilst the melting of the ice so formed is as gradual as the increase of temperature that causes it, the ice of the lakes is more nearly coincident with the calendric than the meteorological winter, or even later than the former. On the 5th of May, 1848, the ice was mostly gone from the *Tso Pangong*, elevated 14,300 feet, but several square miles of it, in a melting and discontinuous state, were still extant in parts of the lake, drifting about before the wind, and the shore was generally margined with large heaps of fragments four inches thick. On the 11th of the same month there were similar remains of ice on the *Tso-Rul*, at 14,400 feet, but they were of small extent and mostly melted when I repassed the lake on the 16th idem. Not a particle of ice was remaining on the *Tso-Moriri*, elevated 15,200 feet, by the 23rd of June, 1849, but the peasants of *Rupshu* reported that it had been frozen over during the past season for five months, viz. December to April (inclusive), and was safe for men and cattle over its whole extent for three of those months, viz. January to March. It must be noted that the first of these lakes is highly charged with salt, and the other two subsaline.

At elevations of 14,000 and 15,000 feet it begins to thaw by day in the end of April and beginning of May; and by the middle or end of the latter month the night frost becomes mild, and the day temperature rises to about 50°.

Regarding the temperature of the highest accessible elevations, my knowledge is confined to a few summer observations in crossing mountain passes; but the following inferences seem probable. At heights between 17,000 and 18,000 feet the temperature rises considerably above the freezing point during the day throughout the summer months: from 18,000 to 19,000 feet it thaws during the afternoon in July and August. At a height of 20,000 or 21,000 feet there is probably "perpetual congelation" in the shade; but the summer sun temperature must still rise above the freezing point to much greater elevations. In these situations the temperature of the air near the surface appears to remain below the freezing point until most of the snow is melted, the heat imparted by the sun being first absorbed in liquefying the snow.

Out of 26 observations, at as many different places, between 17,000 and 19,000 feet, the lowest temperature was 4°, at 11 A.M. 6th of November, 1848, at a height of 17,200 feet, in deep snow; this being the first permanent snow of the season, and the latest date at which I crossed a high snowy pass. In the earliest passage in spring, which was on the 22nd of April, 1848, I found the temperature 8° at 10 A.M., the height being 18,300 feet, and still covered with deep snow: at the latest in spring where I found deep recent snow, it was 12°, at 7 A.M. on the 13th of June, 1849, at a height of 18,800 feet: at the latest in autumn still free from snow, it was 26°, at noon on the 7th of October, 1847, at 18,400 feet. At the greatest attained elevation, which was 19,000 feet, the temperature was 30°, at 9 A.M. on the 11th of July, 1849, a little old snow still lying on the pass-top. The highest observed temperature was 55°, at noon on the 5th of August, 1848; but this was at the lowest elevation, 17,000 feet, and in a sheltered ravine bottom, on the S. side of a pass near 2,000 feet higher.

As I was not provided with proper radiating thermometers, my few Sun observations were not of much value; but one or two of the results may be worth stating. In 46 observations, from May to October, at heights ranging from 11,600 to 18,800 feet, a common thermometer with a plain unblackened bulb, rose when exposed to the sun, from 3° to 42° above the shade temperature shown by the same instrument nearly at the same time. In *Hanle Gunpa*, elevated 14,500 feet, from the 26th of September to the 6th of October 1847, 13 sun observations were made amidst the walls of the buildings, exposed to a great deal of reflected heat and sheltered from the wind, the weather at the time being generally very fine: under these circumstances the extreme observations were—

At 10 A.M. 26th September	Sun	80°	difference 31°
	Shade	49°	
And at noon, 2nd October	Sun	119°	difference 78°
	Shade	41°	

the sun temperature being above 100° in 10 cases out of 13.

In the same place, at noon of the 30th of September, the thermometer inadvertently left against a wall exposed to the sun was found at 145°, the shade temperature at the time being about 37°, and the boiling point of water 186¼°. And again, in the bed of the *Guge Sutluj* under *Tot-Ling*, elevated 12,400 feet, at noon of the 17th of July, 1849, the plain-bulb thermometer, laid upon a white felt on a rocky bank, with a N. aspect, rose in the sun to 140°; whilst the temperature of the air in the shade was 59°, of the river 55°, and of boiling water 190°.

The above, though not showing the true amount of the solar

radiation, sufficiently indicate that it is very great, and the fact is notorious in the personal experience of all travellers in these regions. This heat of the sun's rays compensates, to a certain extent, for the deficiency of atmospheric warmth in these elevated regions, and forms a most important element in the process of vegetation, without which much of the Tibetan agriculture would be impossible.

In the winter of 1848-9 I made an attempt to ascertain the temperature of the earth at the foot of *Le* town, with the idea of obtaining a constant temperature indicative of the mean temperature of the place; but the attempt was unsuccessful, either from want of proper thermometric apparatus, or because it was impossible, with the tools and workmen forthcoming, to attain any depth without making a hole so wide as to admit some of the changes of temperature in the air above. My well was sunk to a depth of 36 feet, and was 3 or 4 feet in diameter; the soil was the common alluvial deposit of granitic gravel, and at the bottom a little moist but not wet. The temperatures at the bottom were—

At 11 A.M. 27th December	1848, 51°,	air above being	24°
At 9 A.M. 28th December	1848, 46°,	do.	24°
At 8 A.M. 24th January	1849, 43°,	do.	18°
At 8 A.M. 25th March	1849, 38°,	do.	34°

which look as though the temperature of the upper air gradually found its way to the bottom of the well during the three months.

I also once or twice noted the temperature of the earth freshly extracted from a narrow bore, which could only be got to the depth of 3 or 4 feet; it was

At 9 A.M. 28th December	1848, from 3 feet, 33°,	air being	24°
At 2½ P.M. 23rd January	1849, from 4 feet, 37°,	do.	30°
At 6 P.M. 24th March	1849, from 3 feet, 40°,	do.	42°

from which it may be inferred, that in these regions the earth is not frozen below the surface, even in mid-winter, in localities so warm as Le, or generally up to heights of 11,000 feet.

In August, 1848, I made similar attempts to bore for subterranean temperatures at a height of 14,800 feet, at *Phoprang* of *Pangong*, but they were totally frustrated by water at a depth of 3 or 4 feet, the site being in a narrow valley-bottom saturated by a large rivulet, and the contiguous hills too stony for the available means of excavation.

In the absence of accurate determinations, some estimate of the mean temperature of Central *Ladak* may be better than nothing: I therefore add the following as probable approximations:—

Central *Ladak*, in lat 34° N. at 12,000 feet	Coldest month, January	from − 5° to + 25°	Mean 10° Fahr.
	Warmest month, July	from + 50° to + 70°	Mean 60° do.
	Mean of the year		35° do.

But the mean temperature of *Le* is probably 38°, having an advantage of site amounting to 2° or 3°, and assimilating its climate to other places less warmly situated at an elevation of 11,000 feet.

The height of freezing mean temperature may be reckoned probably about 13,000 feet.

Barometric Pressure.—The few barometric observations that I was able to make at fixed stations, tend to show that the atmospheric pressure is subject to the same sort of fluctuations on the Tibetan table-land, even at great heights, as prevail in the lower regions elsewhere.

At *Hanle* in the *Changtang* of *Ladak*, at an elevation of 14,400 feet, the daily tides occurred during 17 days of the end of September and beginning of October 1847, about the usual hours, viz., 9 A.M. for the maximum and 4 P.M. for the minimum, and their mean amount during that period was nearly ·09, the average pressure being about 17·8. At a station in the town of *Le*, elevated 12,000 feet, during 14 days of the end of November and beginning of December, 1847, the tides obtained at the same hours and averaged the same amount. In the latter case the pressure fluctuated between 19·31 and 19·77, showing an irregular variation of ·46, or nearly half an inch in a fortnight; but these limits seem to have included the whole amount of the ordinary fluctuation; for during 18 months over which my observations extended at the same place (though by no means continuously), the highest and lowest boiling points of water (out of 17 observations) were 191·45 (at 11 A.M. 24th November, 1848) and 190·30 (at 5 P.M. 17th April, 1848), which corresponded exactly (on the scale of my thermometer) to the barometric extremes above stated.

The irregular variations of pressure seem to be connected in these regions, as elsewhere, with calm and fair weather in the case of high pressure, and wind and moisture when low. My observations were not sufficient to establish any of the regular annual movements of the barometer.

Winds.—In a country formed like West *Nari*, of narrow passages between walls of steep mountain, variously arranged and connected, any single aërial current uniformly impelled from one quarter, must, in all probability, become subdivided into a great variety of local currents, sometimes so altered in force or direction as to be annihilated or diametrically inverted. The complexities of such a system cannot be explained by any off-hand or partial observations of a mere traveller; and what little I say on the subject must be taken with a sufficient margin of doubt.

The total amount of wind in the year is perhaps more moderate than travellers from warmer regions, exposed in tents, are apt to suppose. Winter is the season of least wind. In Central *Ladak* a good deal of high W. wind prevails in spring. Strong S. winds blow during the summer in the southernmost districts of *Ladak* and *Nari-Khorsum*, or those next to the Indian *Himalaya:* they are especially remarkable on the southern confines of *Guge, Kangri* and *Purang*, and in exposed situations often very violent. The force of the wind is greatest in the afternoon, or warmest part of the day, the nights and mornings being usually calm. The prevailing direction of the wind in *Ladak* appears to be W. or N.W., blowing up the valley of the *Indus;* but in open places of the *Changtang* I have sometimes observed a diurnal veering, the wind changing its direction so as to blow from the apparent place of the sun, or easterly in the morning, southerly at noon, and westerly in the evening.

Clouds.—The climate of *Ladak* is sufficiently fine as regards the proportion of sunny to cloudy weather. During two years, I found the number of days on which sunshine predominated, to be nearly as three to two compared with those on which clouds prevailed; the former including perhaps half the number of days (or nearly a third of the whole year) on which the weather was intensely fine, sometimes with a perfectly cloudless sky, and the latter an equal proportion, or about a fifth part of the year, of very dull or gloomy weather. Foul weather, that is actual rain or snow, during the daytime is of very rare occurrence—in the valleys at least, though the mountain tops may experience a good deal more of it.

Winter is the season of clouds; the number of cloudy days in this quarter considerably exceeding the fine ones; in spring they are nearly equal; in summer the sunny weather greatly exceeds the cloudy; and in autumn the sunshine still predominates, in a less degree.

By far the commonest form of cloud is a cirrus, of a hazy and indefinite sort; cumulus may be seen, but rarely, in summer. The nimbus, as seen in the lower regions, is almost unknown here, or faintly approached in midsummer, and seldom perfected to the production of a few drops of rain. The discharge of snow at all other seasons results from a thin misty cloud, formed by insensible gradations from the cirrus, condensed and lowered to the surface of the earth; the descent of such quasi-nimbus into the valleys as low as Central *Ladak*, or 11,000 feet, during the daytime, occurring only twice or thrice in the course of a whole winter, and never at any other season.

The statements here made must not be applied very strictly to the southernmost parts of the country, as they partake, in a slight degree, of the clouds and mists of the Indian *Himalaya*.

Moisture.—I have no wet-bulb or dew-point observations from which to discuss the minutiæ of this subject; but I kept a careful account of all precipitations of moisture susceptible of measurement by rough means, which affords some idea of the extraordinary dryness of the climate of *Ladak*.

It appears from these that, on the average of two years, there were 31 measurable falls of snow in the year, descending to the inhabited valleys at heights of 12,000 feet and upwards, besides 18 falls of snow on the mountain-tops out of reach, and 4 showers of rain too scanty to be measured—in all 53 precipitations of moisture in the year.

The 31 measured falls of snow amounted only to 20¼ inches, and though this amount may have lost a little sometimes by melting before I could effect my measurement, it includes many falls which covered the ground so thinly that they could not be measured (by ordinary means), and were therefore estimated at the twentieth part of an inch each.

The greatest continuous snow was at *Himis Gunpa*, elevated 12,500 feet, from the 23rd to the 25th of April, 1849, and amounted to 8 inches, which in these accounts however I have divided into 5 falls, as it occupied three nights and two days. The annual mean of one fall (within 12 hours, *i. e.*) was only two-thirds of an inch.

The distribution of the snow through the seasons is thus :— Nearly the whole of it in the winter and spring, and this almost equally divided between the two seasons in aggregate quantity; but as regards frequency, there were nearly three falls in the winter to two in the spring, so that the former averaged little more than ½ an inch, while the latter exceeded ¾. In winter the greatest fall in 12 hours was only 2 inches, viz. at *Le*, 12,000 feet, on the night of the 5th-6th of February, 1848; in spring it was 4 inches, at *Erat* of *Tanktse*, 14,000 feet, on the night of the 23rd-24th of May, 1848.

The only summer fall that came under my observation occurred in the earliest part of that season, viz. the 8th and 9th of June, 1849, and in the southernmost part of *Zangskar* close to the Indian *Himalaya*, so that it is rather an anomalous item as regards the climate of Central *Ladak*. Half of all the snow-falls observed on the mountain tops out of reach, occurred in the single season of autumn. In Central *Ladak* little or no snow appears to fall for three months of the year, coincident with the calendric rather than the meteorological summer; and this season may be reckoned safe for crossing the highest passes up to 19,500 feet, but on Indian passes the traveller may be buried in snow, even below 17,000 feet, at the autumnal equinox.

Snow during the daytime is unusual, by far the greater number of falls occurring between midnight and sunrise, or at the coldest part of the 24 hours; a remarkable proof this of excessive dryness.

The rain that came under my observation was confined to the summer; excepting one fall in *Zangskar*, on the very last day of spring (at *sTongde*, 11,900 feet, 31st of May, 1849), and which turned to snow at night; its quantity was contemptible; and, speaking approximately, one may say that it never rains in *Ladak*. Allowing for imperfect measurements, and falls not susceptible of measure, the total amount of moisture precipitated in the year may be reckoned about 2 feet of snow, which is equivalent I believe to only 3 or 4 inches of rain. This refers particularly to elevations about 12,000 feet, and to Central *Ladak:* at superior elevations the fall of snow is probably greater, and I more than once found it lying 2 or 3 feet deep on mountain passes, between 17,000 and 19,000 feet.

In the *Lhogyut* the climate is generally moister, and the quantity of snow very much greater, than in the *Zhunglam;* and the *Changyut* again more dry and snowless than the *Zhunglam*. This is a general effect, depending on proximity to the moisture of India or to the aridity of Central Asia. A local cause of such differences is to be found in the relative height and density of the mountain masses, which are the great collectors of cloud and snow; and in the *Changtang* accordingly there is less snow than in the *Rong*. Hence the excess of snow in *Zangskar* and *Purik*, compared with *Ladak* proper and *Nubra;* and in *Purang* and *Kangri* compared with *Guge* and *Gar;* the moderation of snow in *Rupshu* and in *Guge*, though belonging to the *Lhogyut;* and the snowless waterless sterility of *Ruduk*.

The snow of *Ladak* is remarkable for its form as well as its small quantity. Instead of the large crystallized flakes familiar to English experience, it generally falls in minute granules, somewhat globular, and not obviously crystalline. This seems to be the joint effect of low temperature and rarified atmosphere, and analogous to the fine drizzling rain in the higher regions of the Indian *Himalaya*. The form becomes much more flaky and crystalline in the falls of spring at higher temperatures.

Dew, hoar-frost or frozen dew, and hail are, I believe, unknown, unless in immediate proximity with the Indian *Himalaya*.

Electricity.—The total absence of thunder and lightning in *Ladak* is most remarkable, and appears to be dependent on the excessive dryness of the climate. In two years I twice heard a very faint roll of thunder, accompanied by clouds, and a few drops of rain; and once, close upon the borders of the Indian *Himalaya*, I saw the gleaming of distant lightning, that seemed to belong to some great thunderstorm far on the Indian side of the mountains: all these were in the middle of summer. In winter the increased dryness of the air deprives it of its electrical conduction to such an extent, that the human body, cased in sheep-skins, becomes charged so strongly sometimes, as, on touching a conducting object, to give off large sparks, with a shock that may be felt through the joints of the fingers.

Snow Line.—Immediately connected with the subject of climate is this "vexata questio;" indeed the snow-line of a country may be regarded as an index of its joint thermometric and hygrometric status, and of their mutual reaction. The hygrometric element has often been too much overlooked, partly, no doubt, because it was comparatively unknown; but the preceding observations on the precipitations of moisture in *Ladak* may assist in dispelling these errors of omission, and my brother Richard has already given the necessary theoretical explanations in his memoir on the snow-line of the Indian *Himalaya*. From a series of minute observations on the snow level made during two years, in the course of which I crossed 25 passes, elevated from 15,000 to 19,000 feet, at various seasons, between the end of April and beginning of November, I have arrived at the following conclusions:—

The snow-line in the central and northern parts of West *Nari* attains an extreme height of nearly 20,000 feet. It lowers on approaching the Indian *Himalaya*, and on the southernmost parts of the Indian watershed descends perhaps so low as 18,000 feet. In the *Changtang*, or any districts not in the *Lhogyut* where they are under 20,000 feet, the mountains will be almost entirely denuded of snow during the latter part of summer.

The snow curve appears to culminate about the middle of September, and its season of greatest elevation to be later even than the calendric summer. Its depression by fresh falls in autumn is very gradual in the central and northern zones, so that heights of 18,000 feet may be found free of snow in the middle of October, whereas on the Indian watershed the snow line often descends before the end of September several thousand feet in a few days.

I am not able to define the minimum limit of snow in West *Nari*, because my winter residence lay altogether above it, and out of all sight or cognizance of it, but I doubt whether any part of *Balti*, *i. e.* down to 4,500 feet, can be below the reach of winter snow in ordinary seasons.

It is also probable that the snow which survives the heat of summer above 20,000 feet, does not retain its original form, but is mostly converted to nevée before its renovation by the falls of autumn, and that this effect extends even to the tops of the highest mountains of the central and northern regions.

The snow-line is, of course, generally lower on the N. exposures than on the S. The difference appears to be least, and then hardly appreciable, soon after heavy falls; when the melting has made some progress it sometimes exceeds 1,000 feet.

In these statements I consider the Snow-line as that average level where the surface of snow is equal to the surface denuded by melting (and not merely by precipitous formation of the mountains). The estimates themselves (made from heights determined by the boiling thermometer) are attended with some uncertainty—to the amount perhaps of 500 feet; for though the hypsometric data on which they rest are probably correct within one or two hundred feet, the process of judging by the eye, whether the snow upon one's path, and still more upon the contiguous mountain sides above one's actual station, begins to exceed the bare spaces, or *vice versâ*, is neither easy nor susceptible of much exactness.